The Air Spora

The Air Spora

A manual for catching and identifying airborne biological particles

Maureen E. Lacey and Jonathan S. West

A.C.I.P. Catalogue record for this book is available from the Library of Congress.

ISBN-10 0-387-30252-2(HB)
ISBN-13 978-0-387-30252- 2(HB)
ISBN-10 0-387-30253-0 (e-book)
ISBN-13 978-0-387-30253-9 (e-book)

Published by Springer
P.O. Box 17, 3300 AA Dordrecht, The Netherlands.

www.springer.com

Copyright of the paintings throughout the Work belongs to Maureen E. Lacey.

Printed on acid-free paper

All Rights Reserved
© 2006 Springer
No part of this work may be reproduced, stored in a retrieval system, or transmitted in any form or by any means, electronic, mechanical, photocopying, microfilming, recording or otherwise, without written permission from the Publisher, with the exeption of any material supplied specifically for the purpose of being entered and executed on a computer system, for exclusive use by the purchaser of the work.

Printed in the Netherlands.

Dedicated to
Pauline, Michael and Peter
Oliver, Jonathan and Sophia
and George

Contents

Preface	XIII
Taxonomy and Nomenclature	XI
Chapter 1	1
Introduction to Aerobiology	1
1 What is aerobiology	1
2 The Air Spora	
3 What is in the air	2
3.1 Outdoor air	2
3.2 Indoor air	2
3.3 Biological particles	3
3.4 Inorganic particles	4
4 Early history of aerobiology	4
5 Aerobiology as a discipline	9
5.1 Gregory's basic principals of aerobiology	9
5.2 The Hirst spore trap	10
5.2.1 Early applications of the Hirst trap	11
5.2.2 Manufacturing the Hirst spore trap	11
5.3 Rotating or whirling-arm traps	11
5.4 Recent developments	12
6 Aerobiology in action	13
6.1 British Aerobiology Federation	13
6.2 The National Pollen and Aerobiology Research Unit	13
6.3 Midlands Asthma and Allergy Research Association	14
6.4 International Association for Aerobiology	14
Chapter 2	15
The Aerobiology Pathway	15
1 Introduction	15
2 Take-off (release)	15
2.1 Spore release	16
2.2 Pollen release	17
2.3 Release from lower plants and animals	18

3 Dispersal ... 18
   3.1 Terminal velocity ... 20
   3.2 Aerodynamic diameter ... 20
4 Deposition - sedimentation and impaction ... 21
5 Impact ... 21
   5.1 Plant disease ... 22
      5.1.1 Plant disease symptom distribution ... 22
      5.1.2 Timing of spore releases and disease control ... 22
      5.1.3 Disease gradients ... 23
   5.2 Health Hazards ... 24
      5.2.1 Allergy ... 25
      5.2.2 Late summer asthma ... 27
      5.2.3 Farmer's Lung ... 27
      5.2.4 Other aerobiological hazards in the work place and home ... 28
      5.2.5 Compost handling and locating refuse or composting facilities ... 29
      5.2.6 Respiratory infections ... 29
      5.2.7 Aerobiological hazards in natural environments ... 30
6 Interpreting spore trap data ... 31
   6.1 Clumping of particles ... 31
   6.2 Backtracking wind dispersal ... 31
   6.3 Long distance dispersal ... 32
7 Dispersal by rain splash and aerosol ... 33

Chapter 3 ... 35
Air Sampling Techniques ... 35
1 Introduction ... 35
2 Passive traps ... 36
3 Cascade impactor ... 37
4 Andersen sampler ... 38
5 Whirling arm trap ... 39
   5.1 Description ... 39
   5.2 Siting and running the trap ... 41
   5.3 Using and changing the arms ... 43
   5.4 Mounting the tape strips ... 43
   5.5 Light weight travel trap ... 43
6 Burkard trap ... 43
7 Cyclone and minature cyclone samplers ... 44
8 Virtual impactors and liquid impingers ... 45
9 Filters ... 46

Chapter 4
Using a Burkard Trap

1. Introduction ... 49
   1.1 Safety ... 49
2. Siting the trap ... 49
3. Running the trap ... 50
   3.1 Preparing the drum ... 51
   3.2 Changing the drum in a trap ... 53
   3.3 Mounting the tape from the drum ... 54
4. Using a 24-hour trap ... 57

Chapter 5
Using a microscope

1. Introduction ... 59
2. Microscope structure ... 59
   2.1 Introduction ... 59
   2.2 The eyepiece system ... 59
   2.3 The objectives ... 61
   2.4 The stage ... 61
   2.5 The substage lighting system ... 62
   2.6 Illumination ... 62
   2.7 The stand ... 62
   2.8 Eyepiece graticules ... 63
   2.9 Stage micrometer ... 63
   2.10 Protection ... 63
   2.11 Camera and drawing tube ... 64
3. Using a microscope ... 65
   3.1 Operation ... 65
   3.2 Focusing under high power ... 65
   3.3 Using oil immersion ... 66
   3.4 Practice ... 68
   3.5 Using vernier scales ... 68
   3.6 Ending operations ... 68
4. Measuring and calibration ... 68
   4.1 General remarks on calibration ... 69
   4.2 Calibrating an eyepiece graticule scale ... 70
   4.3 Measuring ... 71
   4.4 Calibrating an eyepiece square graticule ... 71

Chapter 6
Pollen and spore counts 73
1 Introduction 73
    1.1 Examples of slides 75
        Pl. 1. Burkard trap slides taken with x20 objective 75
        Pl. 2. Burkard and travel trap slides taken with x20 objective 77
        Pl. 3. Slides from various traps taken with x20 and x40 objective, health hazards for plants and man 79
    1.2 Choosing the right magnification 80
    1.3 Counting conventions 80
2 Counting pollen and spores from a Burkard trap 81
    2.1 Introduction 81
        2.1.1 Deposition of particles 82
        2.1.2 Daily count - longitudinal traverse 83
        2.1.3 Hourly count - transverse traverses 84
    2.2 The daily pollen and spore count 84
        2.2.1 The positioning of transverse traverses 84
        2.2.2 Counting and recording the daily count 85
        2.2.3 Pollen and spore concentrations and correction factors 86
        2.2.4 Calculating a correction factor 87
3 Counting pollen and spores from a whirling arm trap 89
    3.1 Counting spores on the tape sections 89
    3.2 Calculating particle concentration 89

Chapter 7
Identification 91
1 Introduction 91
2 Particle types 91
    2.1 Pollen 92
    2.2 Fungal spores 92
    2.3 Other plant material 92
    2.4 Animal material 93
    2.5 Inorganic material 93
3 Useful references 93
4 Paintings – x1000 94
    Pl. 4 - Pollen, Monocotyledons and Dicotyledons 97
    Pl. 5 - Pollen, Dicotyledons 99
    Pl. 6 - Pollen, Dicotyledons 101
    Pl. 7 - Coniferous pollen and fern spores 103
    Pl. 8 – Fungal spores - Ascospores and Uredinales 105
    Pl. 9 - Fungal spores - Basidiospores and others 107
    Pl. 10 - Fungal spores - Mitosporic 111
    Pl. 11 – Fungal and moss spores, diatoms and algae 113
    Pl. 12 – Miscellaneous 115

Appendix 117
1 Recipes 117
2 Suppliers 118
3 Web pages 120
4 Counting sheets 120

Glossary 125

References 129

Index 141

Preface

The Air Spora has been studied under this name for over fifty years, providing important scientific advancements particularly in understanding the dispersal, distribution and impact of human, animal and plant pathogens and allergens. One of the most important methods used in studying microscopic particles caught from the air - the Air Spora - is traditional light microscopy, which is the emphasis of this book and a fascinating occupation. Microscopic biological particles commonly found in air include bacteria, spores of fungi and lower plants, pollen, minute animals and debris from life forms. In his book, *The Microbiology of the Atmosphere* Philip Gregory had three colour plates of paintings showing a range of these particles. More than 30 years later, this book revisits the subject and contains 9 plates of paintings showing common and important airborne particles, all to the same scale, to help further in identifying biological particles that may be trapped from the air. It also contains a short history of aerobiology and shows how the subject has developed, particularly in plant pathology, over the last sixty years.

Catching, identifying and quantifying airborne biological particles are important parts of aerobiology. This manual includes a step-by-step guide to key techniques and provides many practical hints for trapping these particles and calculating their concentrations in air. We are indebted to the British Aerobiology Federation (BAF) for giving permission to reproduce many diagrams and instructions from its publication: *Pollens and Spores, A guide to trapping and counting.*

We would like to thank the many people who provided material to help in the preparation of the plates. They are: William Marshall (Antarctica); Peter Burt (Costa Rica); Janiki Bai, Pralip Basu, Pampa Chakraborty, Wadia Kandula, Asha Khandelwal and P.M Reddy (India); Ong Tang Ching (Singapore); M. Trigo Péres (Spain); The late Siwert Nilsson (Sweden); Beverley Adams-Green, John Bailey, Geoffrey Bateman, Colin Campbell, Eric Caulton, Brian Crook, Wendy Milligan, Bob Odle, Judy Pell and Poonam Sharma (UK); Mary Kay O'Rourke (USA); and Ines Hurtado (Venezuela).

We also thank, Peter Burt, Eric Caulton, Julie Corden, Brian Crook and Jean Emberlin for encouragement, helpful comments and contributions, Alastair McCartney for four of the figures, Agneta Burton and Gary Robertson for reading the manuscript. A special thanks goes to Peter Lacey for his constant and patient encouragement, and help in the computing part of the preparation of the plates.

Maureen E. Lacey
Jonathan S. West

Plant Pathogen Interactions
Rothamsted Research
Harpenden
AL5 2JQ
UK

Taxonomy and Nomenclature

Although the authors have attempted to use currently accepted Latin names and classification of organisms, there is generally a great deal of renaming and reclassification of organisms taking place due largely to the use of recently developed molecular biological techniques. As a result, names are subject to change and the authors would welcome any suggestions to correct the taxonomy or nomenclature of organisms mentioned in this book.

Information about any mistakes, omissions, criticisms and suggestions for future improvements would be welcome by contacting the authors at the following address or email and quoting 'The Air Spora':

Dr Jonathan West, Plant Pathogen Interactions, Rothamsted Research, Harpenden, AL5 2JQ, UK

jon.west@bbsrc.ac.uk

CHAPTER 1

Introduction to Aerobiology

1. What is aerobiology

Aerobiology is a study of biological particles present in the air, both outdoors (extramural) and indoors (intramural). Many aspects of our lives are affected by biological particles that are carried in the air and are deposited from it. For example, many people have allergic reactions to inhaled biological particles and many human, animal and plant pathogens are transported by the air. Some organisms are adapted for wind transport whilst others become airborne incidentally, or only as debris. The main incentive for the development of aerobiology as a scientific discipline has been the desire to understand the dispersal of diseases of man, animals and plants in order to try and prevent them. Hence the dispersal of pollen and spores has been the main interest, while the ecology of the air itself has been of secondary importance.

Aerobiology requires an understanding not only of the biological particles being moved by the air but also of physics, as various physical processes explain the movement of air and particles suspended in it.

2. The Air Spora

The term the 'Air Spora' was first used in an article by P.H. Gregory published in *Nature* (Gregory, 1952).

> The population of air-borne particles of plant or animal origin, which will here be called the air 'spora' (taking the Greek σπορά as a word of similar usage to 'flora' and 'fauna'), contains spores and pollens of various shapes ranging in size from 100 μ in diameter for some tree pollens down to 3-5 μ with some of the smallest fungus spores.

In addition to pollen, plant and fungal spores, the air spora may also comprise protists (protozoa), bacteria, viruses and fragments of any biological origin.

3. What is in the air

The air that we breathe not only comprises the gases nitrogen, oxygen and carbon dioxide but also traces of other gases and particles of inorganic and biological material. When there are sufficient contaminants in the air it may be possible to see them with the naked eye in the form of dust, smoke or smog. Inorganic particles can include minute particles of rocks, products of combustion and dust from outer space. Particles of biological origin can include viruses, bacteria, actinomycetes, fragments of fungi and fungal spores, lichen fragments and their spores, protists (e.g. protozoa, algae and diatoms), spores of plants (e.g. mosses and ferns), pollen, plant fragments and small seeds, invertebrates (e.g. nematodes, mites, spiders and insects) and their fragments and faecal material, plus skin, hair, dried mucus and excrement from larger animals. These particles range in size from 1 to >200 µm. Even much larger animals such as frogs, fish and molluscs have been recorded to fall from the sky, presumably following unusual weather events such as tornadoes or waterspouts but these macroscopic organisms, plus flying insects, birds and bats, which can exert at least some control over their flight duration and trajectory, are not considered here in the context of aerobiology. Volatile chemicals in the air, while not living may originate from biological sources as metabolites in processes such as decomposition, fermentation or toxin production.

3.1. *Outdoor air*

A profile of the earth's atmosphere (Fig. 1.1) is shown on a logarithmic altitude scale to enable the various layers to be presented on one page, and to illustrate vividly that the properties of the atmosphere change most sharply near to the ground (Gregory, 1973). Barometric pressure, density of the air, and (as a rule) temperature, decrease with increasing height above the earth's surface. The three vertical panels in Fig. 1.1 show contrasting weather conditions: a still clear night, a cloudy day with increasing wind velocity, and a sunny day. The laminar boundary layer is a still microscopically thin layer of air at the surface of the earth and all objects protruding from it. Above this layer is the variable turbulent boundary layer (planetary boundary or mixing layer) through which most particle dispersion occurs, extending up to the stratosphere (an altitude of 10 km). In addition to turbulence or eddy currents, differential heating from the Earth's surface, particularly on sunny days, leads to pockets of outdoor air that are warmer than surrounding air, and which rise upwards as thermals.

The content of the outdoor air is dynamic, constantly changing with location, weather, season and time of day due to differing effects of these factors on the production, transport and deposition of the various components of the air spora (Mullins 2001).

3.2. *Indoor air*

Over recent years there has been much concern over the 'Indoor Air Quality' of buildings. Many of the indoor pollutants are of biological origin such as bacteria, fungal spo-

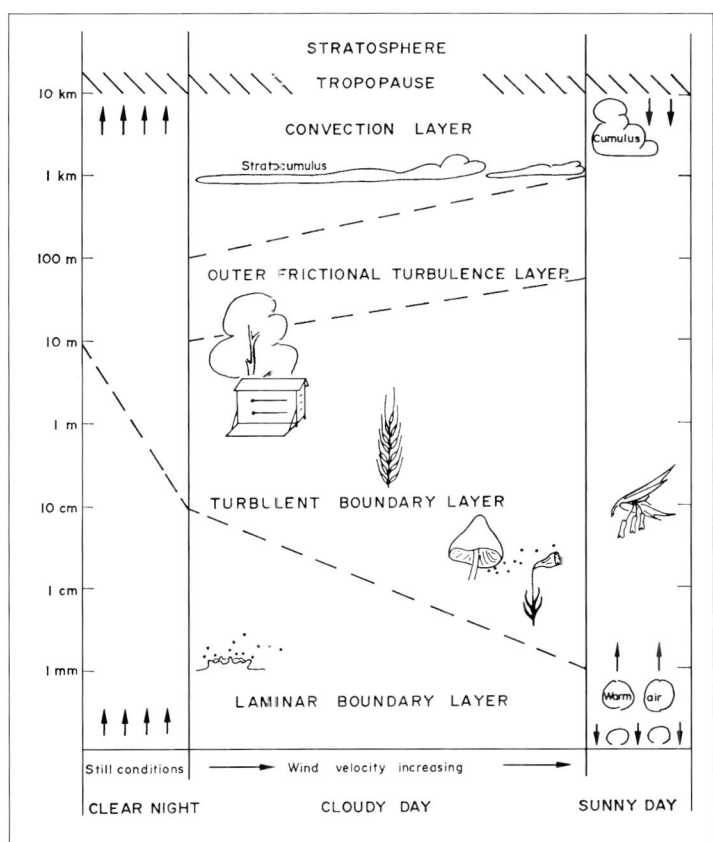

Figure 1.1
Diagrammatic representation of layers of the atmosphere (Gregory 1973).

res (Flannigan, 2001) and also skin, mucus, saliva, nematode eggs and invertebrate faeces. The environmental conditions or microclimate inside buildings is different and less variable than outdoors, leading to an air spora that is relatively homogeneous with time or season, compared to outdoors, but is still very heterogeneous according to the type of building and its use. Clearly, outdoor air is often vented into buildings but the concentration of particular particles will differ indoors compared to in the turbulent outdoor planetary boundary layer, due to local deposition or production of particles indoors.

3.3. *Biological particles*

The majority of pollen in the air comes from the inconspicuous flowers of anemophilous plants (mainly gymnosperms, grasses, and some angiosperm trees), which release clouds of pollen to be blown in the wind. Plants pollinated by insects (or occasionally other organisms such as hummingbirds) have larger, coloured flowers often with nectar to attract the pollinators and the pollen is generally larger, heavier and often sticky.

Viruses, such as the virus causing foot and mouth disease of cattle and sheep, as well as bacteria, algae and protists may become airborne in aerosols produced by splashes of water, urine, wave action or in human or animal breath.

Fungi, actinomycetes, lichens and non-spermatophyte terrestrial plants such as mosses and ferns, reproduce by airborne spores in at least one stage of their life-cycles. Many of these spores are produced and adapted for wind dispersal. A range of mechanisms exist in order for these spores to escape the laminar boundary layer of still air in order for airborne dispersal to be effective (Fig. 2.1).

3.4. *Inorganic particles*

Fine sand or dust has been reported to blow from the Sahara desert in Africa to southern Europe, often reaching as far north as the Alps or southern England before being deposited (usually by rain) (Simons, 1996). This material and also dry clay or other rock dust may become airborne due to eddy currents, whirlwinds or tornados. Additionally, inorganic particles (here not used in the strict Chemistry sense, but meaning particles of non-biological origin) may enter the atmosphere as smoke particles (rather than gasses) from vehicles and fires, and dusts from industrial and other activities of man or from volcanoes. These inorganic particles are concentrated primarily in the layers of the atmosphere closest to the earth's surface, mainly the planetary boundary layer, although smoke from large forest fires can reach several km in altitude and dusts from volcanic eruptions may go higher still, causing light scattering phenomena such as blue moons or halos around the sun (Simons, 1996). Inorganic particles can also enter the earth's atmosphere from space. On average an annual mass of 40,000 tonnes of extraterrestrial dust enters the Earth's atmosphere from space due to the earth's gravity (pers. comm. Matthew Genge, Imperial College, London). These particles reach high temperatures, often melting or partially melting, because although small particles have a relatively slow fall-speed, they are unable to loose heat, produced by air friction, quickly enough.

4. Early History of Aerobiology

In his book *The Microbiology of the Atmosphere*, Philip Gregory (1973) gave a very good historical introduction to the early development of the study of aerobiology, salient sections are quoted in full.

Classical writers believed that the wind sometimes brought sickness to man, animals and crops. Hippocrates… held that men were attacked by epidemic fevers when they inhaled air infected with 'such pollutions as are hostile to the human race'.[1]

Lucretius in about 55 B.C…. observed the scintillation of motes on a sunbeam in a darkened room and concluded that their movement must result from bombardment by innumerable, invisible, moving atoms in the air. This brilliant intuition enabled him to account for many interesting phenomena, including the origin of pestilences.[1]

Following Lucretius, it took over 1500 years before scientists began to realise the diversity of living particles present in air. The belief in 'spontaneous generation' of organisms causing decay and disease was held by many people and persisted for a couple of centuries. Micheli (1679-1737) was a botanist in Florence who, by putting spores of moulds on slices of fruit, showed that they were 'seeds' of the moulds. As some control slices became contaminated he concluded that spores of moulds were distributed through the air (Buller, 1915). In his letters to the Royal Society in 1680 Anton van Leeuwenhoek reported that he was able to see minute organisms with his handmade lenses, he later came to suppose that 'animalcules could be carried over by the wind, along with the bits of dust floating in the air' (Dobell, 1932).

J.G. Koelrueter, in 1766, was perhaps the first to recognize the importance of wind pollination for some plants and of insect pollination for others. C.P. Sprengel in 1793 developed these views and concluded that flowers lacking a corolla are usually pollinated in a mechanical fashion by wind. Such flowers have to produce large quantities of light and easily-transported pollen, much of which misses its target or is washed out of the air by rain. T.A. Knight in 1799 reported that wind could transport pollen to great distances.[1]

By the beginning of the nineteenth century, therefore, it was recognised that pollen of many, but by no means all, species of flowering plants, and the microscopic spores of ferns, mosses, and fungi – as well as protista [protozoa] – were commonly liberated into the air and transported by the wind. The potential sources of the air spora had been discovered and identified in the main before the year 1800, but their role remained obscure.[1]

Ehrenberg accumulated evidence that small microscopic particles might be carried great distances by wind in a viable state. Ehrenberg found sixty-seven kinds of organisms in dust (Fig. 1.2) collected by Charles Darwin (near the Cape Verde Islands in 1833 during his voyage on the Beagle. Darwin had found the atmosphere hazy with dust from Africa and realised at once the importance of Ehrenberg's findings to the geographical distribution of organisms (Darwin, 1846). There are many diatoms in the illustration indicating proximity to water.

Louis Pasteur (1822-95) worked for many years on the cause of putrefaction. He sterilised flasks containing nutrient medium and exposed them to air in different situations, disproving the idea of spontaneous generation and showing that infection was caused by germs (Pasteur, 1861). He also developed a gun-cotton filter to extract suspended dust from the air for microscopic examination.

Pasteur had demonstrated visually the existence of an air spora, he had pointed out that it should be measured while in suspension and not after deposition on surfaces, and he made the first rough visual measurement of its concentration in the atmosphere of the city of Paris: a few metres above the ground in the Rue d'Ulm, after a succession of fine days in summer, several thousands of microorganisms were carried in suspension per cubic metre of air. He then abandoned the method – remarking, however, that it could doubtless be improved and used more extensively to study the effects of seasons and localities, and especially during outbreaks of infectious diseases.[1]

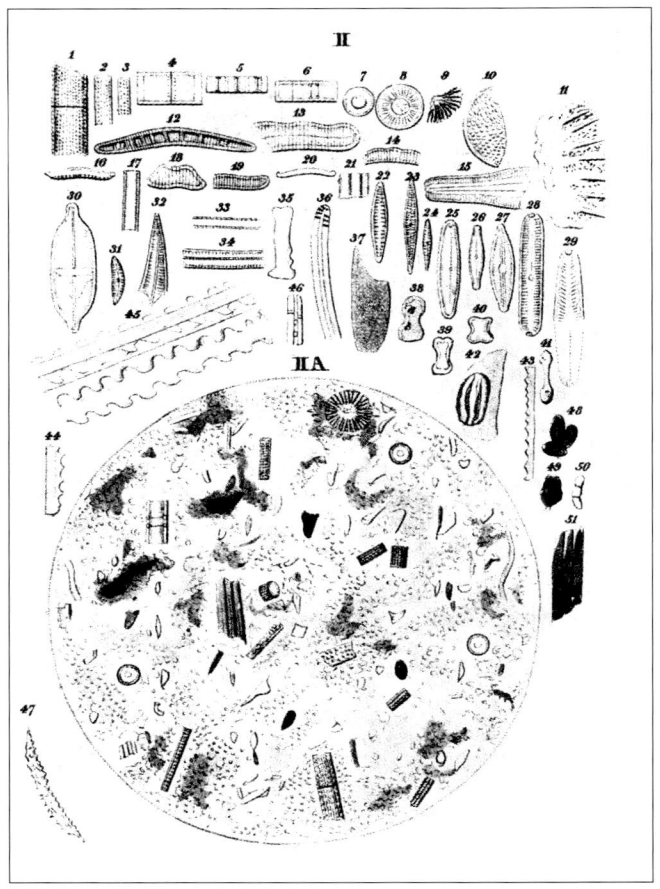

Figure 1.2
Ehrenberg's illustration of dust collected by Charles Darwin on the Beagle near the Cape Verde Islands, January 1833 (Gregory, 1973).

During the last half of the 19th century bacteriological work in laboratories and clinics identified the causes of disease in man (see Bulloch, 1938); e.g. Robert Koch identified the cause of anthrax, tuberculosis and cholera between 1876 and 1883. His statement of Koch's postulates has given a method of confirming that a disease is caused by a particular organism. The suspect is isolated, inoculated into a healthy specimen of the plant or animal, and if the disease develops and can then be re-isolated, it is proved to be the cause of the disease (Holliday, 1992).

Other workers investigated outdoor air to see if the microbes present were connected to disease. Maddox (1870) invented the 'aeroconiscope' and Cunningham (1873) developed this to use in two gaols in Calcutta where cholera and fevers were rife. His aeroconiscope (Fig. 1.3) consisted of a conical funnel with the mouth directed into the wind by a vane and ending in a nozzle behind which dust from the air was impacted on a sticky microscope cover glass. He sampled for 24-hour periods but no correlation was found between his catch and the diseases. His catches were mainly of fungal spores and pollen (Fig. 1.4).

The Observatoire Montsouris, situated south of Paris, was set up in 1871 to make records needed for meteorology and agriculture. The dust in the air was also studied. P. Miquel was the first to make a long-term survey of the microbial content of the atmos-

Figure 1.3
Cunningham's aeroconiscope. A = side view of apparatus (partly in section); B = face of sticky surface behind apex of cone (on larger scale) (Gregory, 1973).

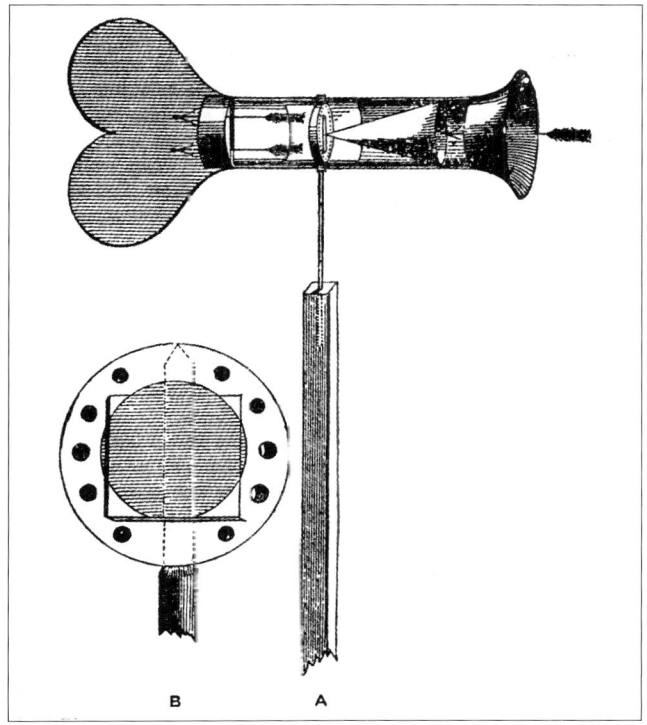

Figure 1.4
Spores collected in a Calcutta gaol, 1872 (Cunningham, 1873).

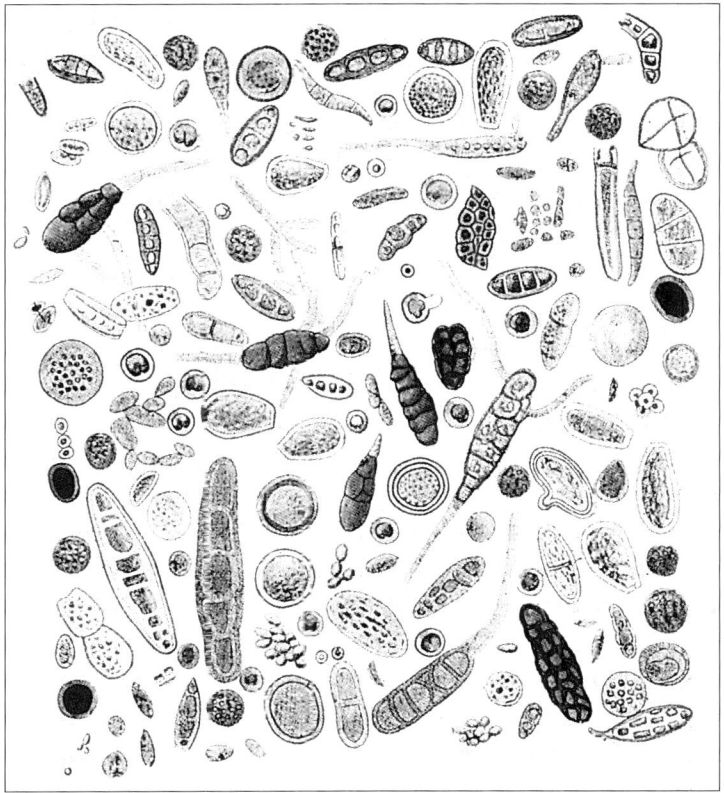

INTRODUCTION TO AEROBIOLOGY

phere by volumetric methods. He used a water operated pump to produce a suction of 20 litres of air per hour through an orifice to impinge on a glycerine-coated glass slide. His estimates of mould spores outdoors in the Parc Montsouris averaged about 30,000 per m^3 in summer, occasionally up to 200,000 in rainy weather. Numbers of airborne bacteria were high in the centre of Paris compared to the park, but higher still in dwellings and particularly in the crowded hospitals (Miquel, 1883). As it became clear that many epidemic diseases were caused by bacteria, work at the laboratory concentrated on the bacterial analysis of drinking water.

In Germany, Hesse made an apparatus consisting of a narrow horizontal tube containing a layer of Koch's nutrient gelatine. Air was aspirated slowly through the tube allowing microorganisms in the air to settle and grow on the medium.

Hesse found that moulds penetrated much further into the tube than did the bacteria, and he made the important deduction that mould germs as found in the atmosphere are on average lighter than the bacterial germs. This led him to conclude that, whereas fungal spores were usually present in air as single particles, the aerial bacteria mostly occur in the air as large aggregates or attached to relatively large carrier particles of dust, soil or debris (Hesse, 1884; 1888). He also observed that most colonies consisted of single species – bacteria usually in small colonies of pure culture and fungi as isolated spores – and deduced that the airborne germs are not in the form of aggregates of different species.[1]

In London, P.F. Frankland (1886, 1887) used Hesse's method to study the air on the roof of Imperial College, London and inside crowded or empty buildings. He also used horizontal dishes with Koch's nutrient medium. He noticed that the number of colonies was greater when the mouth of the tube faced the wind rather than in other directions.

Frankland seems to have been the first person to realize that aerodynamic effects are of major importance in techniques for trapping the air spora.[1]

Although hay fever had been attributed to inhalation of pollen it was not until Blackley (1873) did inhalation experiments on himself that this was proved to be correct. Inhaled fungal spores were also recognised as allergens by Cadham (1924) and Feinberg (1935). Stepanov (1935) was one of the first to try to understand the processes of dispersal of spores.

By the early years of this [the 20th] century it became possible to assess the value of the ancient belief that the wind brings disease. Many diseases of crops, but very few diseases of man, have proved to be caused by minute particles carried on the wind. The particles are not some sort of invisible atoms as Lucretius thought; indeed, among the motes in the sunbeam, he may himself have been watching some of the baleful fungal spores and pollens which cause crop diseases and respiratory allergy.[1]

---

1   From Gregory, 1973.

5. Aerobiology as a discipline

In the 1930s the American F.C. Meier first used the word **Aerobiology** to describe a research project on microbial life in the upper air. Unfortunately he was killed in an air accident over the Pacific Ocean in 1938 when only preliminary abstracts of his work had been published (Haskell and Barss, 1939). The new discipline was eventually launched by a symposium on extramural and intramural aerobiology published by the American Association for the Advancement of Science (Moulton, 1942).

Scientists at Rothamsted Research have made a major contribution to aerobiology over the past 60 years primarily by studying the epidemiology of plant diseases (Hirst, 1994). Philip Gregory has been called the father of modern aerobiology and it was his inspiration that initiated work on air sampling resolving many basic principles (Hirst, 1990, 1992; Lacey *et al.*, 1997). The different stages of the aerobiology pathway, (Fig.1.6. and Chapter 2) have been studied as ways of understanding, forecasting, controlling and preventing the spread of plant diseases. His interest in medical mycology started when he went to Winnipeg, Canada to work on human pathogenic fungi under the guidance of the mycologist, A.H.R. Buller. Because of the great economic depression he had to return to England in 1934 and was able to return to work on the diseases of narcissi at Seale Hayne College in Devon. With food shortages at the start of the Second World War he went to work on virus diseases of potatoes at Rothamsted Research Station.

5.1 *Gregory's basic principals of aerobiology*

Gregory observed infection gradients while working on insect-vectored virus diseases of potatoes (Gregory and Read, 1949). His interests turned to fungal spores and he read widely on the subject, even learning Russian to translate a 1935 paper by Stepanov. This lead to Gregory's paper on 'The dispersion of airborne spores' (Gregory, 1945), which demonstrated a clear understanding of the physical factors controlling the dispersal of both single and clumps of spores (and pollen). He developed and tested his theories, comparing them with others such as Stepanov (1935) and the meteorologists, Schmidt (1925) and Sutton (1932). He discussed the terminal velocity of spores, eddy diffusion, dispersion from both point and line sources, transport by wind and deposition. He observed that gradients of airborne plant infections originating from a point source were closely predicted by his theory but those known, or suspected of being splash dispersed, were not.

Some of Gregory's early experiments used *Lycopodium* spores, liberated after working hours into the natural draught along the corridor of the Plant Pathology (North) building at Rothamsted, and trapped on sticky slides and cylinders, in Petri dishes and, as a volumetric standard, a cascade impactor (Hirst, 1990, 1992). The results showed that many careful experiments were needed to explain how particle size, wind speed, turbulence and the dimensions and configuration of the trap surfaces affected deposition (see Chapter 2). These physical properties were studied, using a purpose-built small wind tunnel, which had a 30 cm (1 ft) cross section (Fig. 2.7) and enabled wind speeds to be varied from 0 to 10 m s^{-1} (Gregory, 1951; Gregory and Stedman, 1953). The work was continued by O.J. Stedman, J.M. Hirst and F. Last (Gregory moved to Imperial College as Professor of Botany in 1954), establishing the standard measurement of air spora as the number of spores m^{-3} air.

5.2 The Hirst spore trap

Jim Hirst (Bainbridge and Brent, 1999) worked at Rothamsted, initially on potato blight, and realised that a reliable suction trap was needed to sample the air for plant pathogen spores rather than rely on the available traps (sticky cylinders). The cascade impactor (May, 1945 and Fig. 3.2) was used as the standard in calibrating other traps but the surfaces onto which particles were deposited became overloaded very quickly. Hirst resolved to use just one orifice (the 2nd; 2 mm wide), but moved a sticky slide past this using a mechanical clock (Hirst, 1952). The resulting deposit was more countable and the time of deposition could be calculated enabling diurnal periodicity or association with meteorological events to be established (traps are normally changed at 9.00 a.m. to coincide with weather records). The suction speed was set at 10 l min to give isokinetic efficiency at mean outdoor wind speeds, a wind vane enabled the orifice to point into the wind.

The Hirst spore trap (Fig. 1.5) revealed a wide diversity of air spora, mainly comprising pollen grains and spores of *Cladosporium, Alternaria,* smuts and rusts in dry wea-

Figure 1.5
A Hirst 24-hour volumetric spore trap (Lacey, J., 1996, with permission from *Mycological Research*).

ther, while at night and after rain there were many hyaline spores including ascospores and basidiospores (Hirst, 1953). Initially (summer 1952), many spores trapped were not identified and could only be placed in 'broad-form' groups e.g. 'dark basidiospore' (Gregory and Hirst, 1957). Other early studies of the total air spora using the Hirst trap were at an estuary (Gregory and Sreeramulu, 1958), at two contrasting rural sites (Lacey, M., 1962) and a two-year study over a paddy field in India (Sreeramulu and Ramalingham, 1966).

5.2.1 Early applications of the Hirst trap

Gregory was an asthma sufferer, who had worked in medical mycology, and thought that the occurrence of basidiospores in the air might be related to symptoms of some sufferers. Counts were made of hyaline, yellow and dark basidiospores during the months of August and September of 1951 with the suggestion that the spores could be allergenic (Gregory and Hirst, 1952). Two buildings containing fructifying dry rot fungus, (then named *Merulius lacrymans*, now *Serpula lacrymans*, Pl. 9.29), were also sampled (Gregory *et al.*, 1953). Thus the Hirst spore trap became an established air sampling technique for health (both indoor and outdoor) and plant pathological studies.

5.2.2 Manufacturing the Hirst spore trap

The great potential of the Hirst trap soon became widely appreciated, particularly among the medical profession. In 1953 Casella Ltd commercially manufactured the Hirst trap and it was operated at St Mary's Hospital, Paddington (Hamilton, 1959) and Cardiff (Hyde, 1959; Hyde and Adams, 1960). The design was improved and in 1966 Burkard Manufacturing Co. Ltd started to produce the seven-day recording volumetric spore trap. A battery powered pump and a 24-hour single slide holder were further developements. These traps, and the Lanzoni VPPS Hirst-type trap are now used in many countries for monitoring the air for airborne biological particles including plant pathogens, and allergens, and provide samples for the pollen counts given with the weather forecasts (see Chapter 4 for details of operating the Burkard trap).

5.3 *Rotating or whirling arm traps*

The development of a trap based on impaction of particles on sticky arms rotating through the air, powered by a simple electric motor, was another important advancement (Perkins, 1957). Due to its relatively low cost and compact size, numerous rotating arm (or rotorod) traps can be used collectively, improving the quality of information gained in studies on the distribution of airborne particles at different heights, directions and distances around a source (see Chapter 3). For example, rotating arm traps were used to assess pollen dispersal around a non-GM maize crop in France to assess risk of GM pollen dispersal (Jarosz *et al.*, 2003).

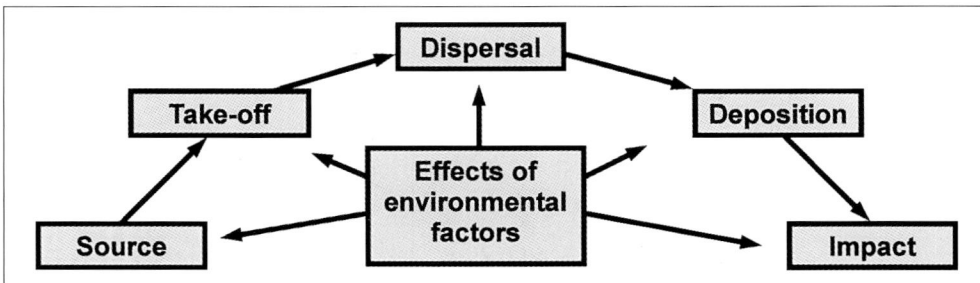

Figure 1.6 The aerobiology pathway (Lacey, J., 1996, with permission from *Mycological Research*).

5.4 Recent developments

Americans Edmonds and Benninghoff (1973) first published the concept of the **Aerobiology Pathway** (Fig. 1.6) as a simple method of explaining the different stages of the transport of organisms through the air. This concept was established and expounded by Edmonds (1979) and Cox (1987). The processes include the production and release of the biological particle, its dispersal through the air, its deposition and impact (effect) on the substrate on which it lands. Much of the early work was done outdoors, but with current emphasis on health and safety, much research is now focussed on indoor environments (Flannigan *et al.*, 2001).

Furthermore, recent developments have tended to reduce emphasis on visual identification of airborne particles in favour of more automated methods of detection and quantification. These techniques will be discussed here only briefly. Initial trapping of airborne particles often uses well-established principles but identification methods have diversified considerably. The integration of air-sampling methods with molecular and immunological diagnostic techniques for example can avoid the tedious nature of particle identification and even quantification (Williams *et al.*, 2001; Calderon *et al.*, 2002; Fraanije et al., 2005). In particular, molecular techniques, DNA or RNA probe technology and polymerase chain reaction (PCR), can be applied to air samples and may answer many previously unanswered or unasked questions such as proof of individual clones of plant pathogens being dispersed large distances to infect crops in different countries or even different continents (Hovmøller *et al.*, 2002; Brown and Hovmøller, 2002). The technique developed by Calderon *et al.* (2002), detected spores collected on the surface of waxed tape (from Burkard or rotating arm samplers) by probing for specific target DNA. Williams *et al.* (2001) described a technique that uses microscopic glass beads (Ballotini beads), in a shaker to disrupt spores (of *Penicillium roqueforti*), collected directly in Eppendorf tubes using a miniature cyclone sampler, and followed by detection using PCR. Air sampling coupled with molecular techniques has also proven to be a very convenient way of assessing the genetic diversity of a population of the target spore-producing organism in a particular region, this has also been helped by the development of the miniature cyclone sampler. However, molecular techniques are not always necessary in diversity studies; Limpert *et al.* (1999) used a jet spore sampler mounted on a car, which was driven across Europe to sample the diversity of barley powdery mildew virulence. Spores were deposited into Petri dishes containing detached barley leaf sections, with virulence of spores from the resulting colonies tested on a differential set of cultivars.

The use of immunological techniques in aerobiology also has been facilitated particularly through development of the miniature cyclone sampler (Emberlin and Baboonian, 1995), a rotating arm sampler modified for collecting spores in wells of rows of a microtitre plate (Schmechel et al., 1996), and a microtiter immunospore trapping device (MTIST) (Kennedy et al., 2000). The latter technique uses a suction system to trap air particulates by impaction directly in microtiter wells, enabling detection and quantification of target particulates by enzyme-linked immunosorbent assay (ELISA).

Another (automated) development identifies target particles e.g. spores of a particular fungus using flow cytometry [of impinged samples] (possibly enhanced with selective staining and use of image analysis and neural network systems) (Day et al., 2002; Morris et al. 1992). Flow cytometry allows spores and other particles (which would need to be trapped from the air and incorporated into liquid) to be analysed optically. For each particle, several parameters are analysed (e.g. light scatter, autofluorescence, particle width) to allow discrimination.

Improvements in modern computing power allowed the development of laser induced fluorescence spectroscopy, which can be used to detect bioaerosols. The apparatus of Cheng et al. (1999) found four bacteria tested to have similar fluorescence spectra, while Eversole et al. (2001) developed a prototype single particle fluorescence analyser, which could detect concentrations of bioaerosols as low as a few (1-5) particles per litre.

The Biotrace Biological Detection System (B3DS) and the Biotrace Intelligent Cyclone Air Sampler (ICAS) (Biotrace International plc.) use wet cyclones to trap airborne particles into liquid, which is then processed in a continuous flow luminometer to give near real-time detection of microbial contamination by ATP bioluminescence.

Filters can be used to detect toxins present in the air or airborne particles such as fungal spores following deposition e.g. onto porous polycarbonate or cellulose filters, followed by appropriate extraction and purification techniques for toxins (GC-MS, HPLC, TLC etc) or followed by diagnosis/culturing of plant or fungal spores or bacteria (Skaug et al., 2001; Agranovski et al. 2002).

6. Aerobiology in action

6.1 *British Aerobiology Federation*

The British Aerobiology Federation was formed in 1990 to bring together people interested in aerobiology in the UK and to promote work and research in the subject area. BAF holds regular scientific meetings and workshops.

6.2 *The National Pollen and Aerobiology Research Unit*

The National Pollen and Aerobiology Research Unit conducts research on aerobiology and in particular the abundance and dispersal of pollen. The work covers many aspects including distribution patterns in allergenic pollen, pollen monitoring, fever and asthma, testing filters, forensics and the dispersal of pollen from GM crops. The UK pollen monitoring network has 33 sites, monitoring seasonal changes and the geographic distribution of pollen. Thirteen of these sites monitor the major allergenic pollen (grass and tree pollen), while others monitor grass pollen during the peak season (early summer) only.

6.3 *Midlands Asthma and Allergy Research Association*

The Midlands Asthma and Allergy Research Association (MAARA) is a charity which conducts and funds research into asthma and other allergies. MAARA has carried out aerobiological research since the charity was founded in 1968 and has the longest pollen and fungal spore dataset in the UK and one of the longest in the world.

6.4 *International Association for Aerobiology*

The International Association for Aerobiology (IAA) was founded at a meeting at the 1st International Congress of Ecology at The Hague on October 11th 1974. The IAA organises the Quadrennial Congress (International Congress on Aerobiology - ICA) which includes plenary sessions, symposia, scientific meetings, meetings of sections, commissions, committees, working groups and exhibitions on all aspects of aerobiology.

CHAPTER 2

The Aerobiology Pathway

1. Introduction

The aerobiology pathway shown in Fig. 1.6 gives the different stages in the movement of particles such as spores or pollen from their source to the effect they cause when they land. A combination of more than one process is often studied rather than each process individually. Sampling the air for certain particles (usually spores or pollens) can be useful for monitoring climate change, estimating or forecasting dispersal of pathogens or allergens or species colonising new habitats, genetic diversity of a spore-producing organism, detection of pathogens or allergens, and risk assessment of GM pollen spread or cross-pollination of plant varieties. For example, air sampling is a valuable tool in the study of crop disease epidemiology and has enabled a better understanding of many crop diseases, leading to disease forecasting, changes in cropping practice to escape disease and optimised fungicide use. It has also enabled sources of hay fever to be identified and warnings of hay fever given during the year based on allergen detection. Indoors it has identified sources of microbial contamination in medical and food processing situations. This chapter considers the different processes in the aerobiology pathway and explains ways to interpret spore trap data. Since much of the early work was done by Gregory and his collaborators, many of the examples given below are from work at Rothamsted Research (Hirst, 1994).

2. Take-off (release)

Aerobiological particles can be considered to originate from point, line or area sources depending on the scale under consideration or type of sampling used. As microorganisms may be widespread while others are confined to rare niche microenvironments, this affects the numbers and distribution of their propagules. Bacteria generally lack mechanisms to become airborne, occurring opportunistically in aerosol generated by rain splash, bubble burst, animal or mechanical activity, and often as aggregations of many viable units on plant and animal debris. Viruses, like bacteria, often become airborne opportunistically from animal, fungal or plant sources and usually as aggregations of particles.

2.1. Spore release

Fungal spores however, vary greatly in size, shape, colour and method of release (Ingold, 1971). These various release mechanisms are essential for spores to escape the laminar boundary layer of still air to be dispersed in the turbulent boundary layer Fig. 1.1, (Gregory, 1973). Passive release of spores occurs, particularly in fungi growing on raised structures e.g. powdery mildew growing on plant leaves, where gusts of turbulence can penetrate closely enough to the substrate to detach spores. This is assisted in the case of powdery mildew by basipetal spore production, the oldest spores being raised away from the leaf on chains of progressively produced spores. However, many fungi have evolved active methods of spore liberation, some of which are illustrated in Fig. 2.1 (see also, Ingold, 1999).

The concentration of some dry airborne spores, e.g. *Cladosporium*, can increase at the start of rainfall. Hirst and Stedman (1963) showed that both rapid air movement in advance of splashes and vibration can blow or tap spores into the air. This process is most effective when large drops collide with surfaces carrying spores that are loose or raised above the surface and is different from true rain-splash dispersal in which spores mix with the water rather than remaining dry (section 7, this chapter).

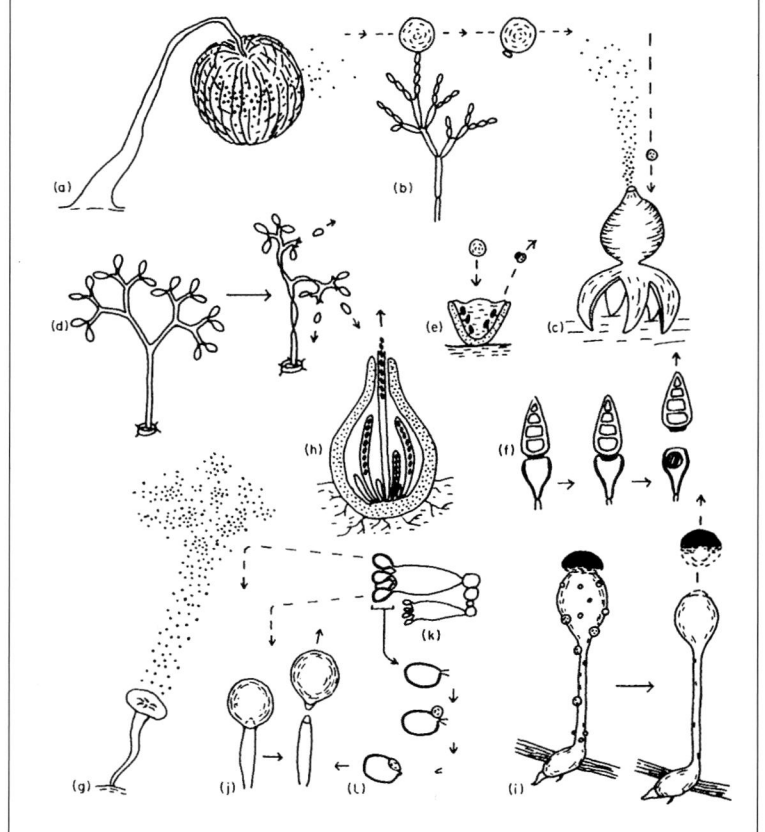

Figure 2.1
Spore liberation mechanisms: (a) deflation from raised fruiting body of *Dictydium* sp., (b) mist pick-up of *Cladosporium* sp. (Pl. 11.1), (c) bellows mechanism in *Geastrium* sp., (d) hygroscopic movements in *Peronospora* sp. (Pl. 9.68), (e) splash cup in *Crucibulum vulgare*, (f) water rupture in *Deightoniella torulosa*, (g) squirt gun (discomycete type) in *Sclerotinia sclerotiorum* (Pl. 8.19), (h) squirt gun (Pyrenomycete type) in *Sordaria fimicola* (Pl. 8.28), (i) squirting mechanism in *Pilobolus kleinii*, (j) rounding of turgid cells in *Entomophthora* sp. (Pl. 9.73), (k. l) ballistospore discharge in *Agaricus* sp. (Pl. 9.1). (Lacey, J., 1996, with permission from *Mycological Research*).

Spores of certain fungi are released seasonally rather than throughout the year and the timing of spore release can be monitored and ideally predicted if a good understanding of climatic effects on fruiting body development and sporulation is established. Often seasonal release of fungal plant pathogen spores is synchronised finely to coincide with a particular growth stage of the host plant e.g. spores of *Claviceps purpurea* (Pl. 8.15), (which causes ergot of cereals and grasses) and *Venturia inaequalis* (Pl. 8. 16), (which causes apple scab) are both released around the time of flowering of their hosts. Studies of spores by Last (1955) within wheat and barley crops infected by mildew *Blumeria graminis (Erysiphe graminis)*, revealed a daily periodicity in spore release. Air at different levels above the ground and at different times of the day was sampled with a portable, manually operated volumetric spore trap (Gregory, 1954). The most abundant fungal spores in the air on a dry day were *Blumeria*, (Pl. 10.30), *Cladosporium* (Pl. 11.1-2), and *Alternaria* (Pl. 11.3-6), with a peak in numbers at 16.00 GMT, while *Sporobolomyces* (Pl. 10.5) and *Tilletiopsis* (Pl. 10.4) were most numerous at 04.00 GMT. After rain, as well as *Sporobolomyces* and *Teletiopsis*, there were many spores tentatively identified as ascospores. It is thought that these periodic differences in the air spora profile reflect different mechanisms of spore release with maximal numbers of dry-spores and pollen released in late afternoon.

Other studies have since confirmed that ascospores are usually released after wetting by rain or dew, the water creating turgor pressure to force the ascospores from the ascus individually in some species or otherwise in one go. Although associated with rain, spore trapping experiments showed that most ascospores were released after rainfall, for up to five days, while the crop debris bearing apothecia was still wet. 'Leaf' wetness and 'Debris' wetness sensors were used to monitor the crop and debris. Tests in a miniature wind tunnel showed that under wet-dry cycles, spores could be produced for as long as 21 days, the largest numbers ejected whilst the debris was drying (McCartney and Lacey, M., 1990). Similarly, ascospores of *Leptosphaeria maculans* (phoma stem canker, Pl. 8.3) were released after rain, and on wet debris had a diurnal periodicity (possibly due to changes in relative humidity), most spores being released around 10 am -12 midday (West *et al.* 2002a).

2.2. *Pollen release*

In gymnosperms and angiosperms, pollen release is passive, with the flower parts raised into more turbulent air and anthers of anemophilous angiosperms often extended on long filaments into the airflow. Compared to insect-pollinated plants, large numbers of relatively small pollen grains are produced by anemophilous plants to ensure that some pollen will reach the intended target. Pollen from plants considered to be insectpollinated can in some cases become airborne, e.g. oilseed rape pollen and may lead to allergy problems, but generally, pollen of entomophilous plants represent a low proportion of airborne pollen e.g. <2% of pollen trapped in Cardiff (Mullins and Emberlin, 1997).

As with the release of spores of many fungal species, the release of pollen is seasonal (Fig. 2.6) and varies according to species and geographical location. The source strength for a particular species varies regionally due to differences in habitat and timing of flowering (Spieksma *et al.*, 2003). This can be seen by clear differences in the start of the grass pollen season throughout the UK (Emberlin *et al.* 1994), regional variations in *Betula* pollen production in the UK (Corden *et al.*, 2000), the grass pollen season in the UK and Spain (Sánchez Mesa *et al.*, 2003) or distribution of Japanese Cedar pollen production (Kawashima and Takahashi, 1999)

Pollen production by crops also varies considerably with time of day, stage of flowering and weather events. Some crops produce large quantities of airborne pollen e.g. above a sugar beet crop, the maximum daily average concentrations reported was 12400 m^{-3}, while for oilseed rape it was 5295 m^{-3} (Scott, 1970). Free *et al.* (1975) measured maximum hourly counts at 46714 pollen grains m^{-3} of air for sugar beet and 2273 m^{-3} for oilseed rape crops. However, these measurements really include a component of dispersal with particle numbers decreasing with height above the source.

2.3. *Release from lower plants, animals, etc.*

Algae and diatoms can become airborne via sea-foam and bursting bubbles (Schlichting, 1971; 1974) and aerosol formation by waves, rough water (rapids, waterfalls, etc). In some mosses such as *Sphagnum*, release of spores is explosive, as drying of the spore capsule increases the internal air pressure until an operculum in the top of the capsule ruptures. Similarly spores of many homosporous ferns are released actively following dehiscence of sporangia which curve back on themselves due to an annulus of thickened cells, but spring forwards again, releasing spores as water in the annular cells turns to vapour (Ingold, 1939). In the horsetails (*Equisetum*, Pl. 7.12) spores are wrapped by four arms (part of the spore coat), called elaters, which in dry conditions, spring open to assist spore release.

In the animal world, protozoa, nematodes, mites and small insects can become airborne by wind action on water, soil, or plants or by mechanical activity (rain-splash, shaking clothing, etc). A special case is certain spiders, which 'balloon' by deliberately extending relatively long silk filaments to catch on the wind (Weyman *et al.*, 2002).

3. Dispersal

Once particles have been launched into the air they disperse, their concentration per unit volume of air decreasing with increasing distance from the point of liberation (Gregory, 1973). This is illustrated by the appearance of smoke billowing from a chimney, which disperses, often showing effects of air turbulence (Fig. 2.2). Expansion of the cloud of particles occurs due to eddy currents, causing dilution of the particle cloud as it moves in the general wind direction.

Dispersal within and above crops is difficult to measure alone as air movement affects the release, dispersal and deposition of fungal spores (Legg and Bainbridge, 1978; Legg, 1983) and pollen. Gust penetration into crop canopies is important for liberation and deposition of spores (Aylor *et al.*, 1981; Shaw and McCartney, 1985), an important consideration in the development of a spore dispersal model (McCartney

Figure 2.2
Smoke dispersal from a chimney in Calcutta, 1997.

and Fitt, 1985; Fitt and McCartney, 1986). However, it also increases dispersal. Last (1955) showed that when spores were formed in the crop the spore concentration was always greater within than above the crop, and also near the ground than at the top of the crop. This is not only due to dispersal but also due to deposition on leaves by the filtering effect of the crop canopy.

Particle dispersal is largely dependent on air mass movement, turbulence and thermal convection. Characteristics of particles such as size, shape, density and surface texture affect dispersal only very subtly, by affecting aerodynamics such as the particle's terminal velocity.

Recently, attention to the dispersal of pollen has heightened due to concerns over the possible spread of genetically modified material. Prior to the development of GM crops, as now, information on pollen dispersal was important to calculate suitable separation distances for seed-production crops so that cross-pollination is minimised. For sugar beet, Chamberlain (1967) suggested that the then recommended minimum spacing of 1000 m from a 20 acre (8.1 hectare) source to a seed-production plot, would result in the proportion of cross pollination to within-plot pollination to be 4×10^{-3} with 1×10^{-3} pollinated from the regional background (long distance) pollen. He suggested that increasing the separation distance to 2000 m would reduce the proportion of cross pollination from the source area to that of cross pollination from the background pollen. The concentration of pollen or other particles is affected by the height above ground (dispersal from the source). Hart *et al.* (1994) described concentrations of grass and nettle pollen trapped using Burkard traps simultaneously at three heights (12, 24 and 30 m) at Leicester, England. They found that pollen concentrations were generally (but not always) lower in the 30 m trap and this was thought to be due to locally produced pollens not mixing enough to reach 30 m. Peaks of pollen grains trapped were later for the higher traps than the 12m trap and this could represent pollen production from distant sources rather than local sources. McCartney and Lacey, M. (1991b) also found a decrease in pollen numbers with height above the source and predicted that more than 60% of oilseed rape (*Brassica napus* Pl. 5.19) pollen lost from a crop would still be airborne at 100 m downwind, but that the concentration at the ground (i.e. available for pollinating a neighbouring crop) would be between 2 and 10% of that at the edge of the crop. Similarly, Jarosz *et al.* (2003) have reported the dispersal gradient of conventional maize pollen, which produced between 2×10^4 and 2×10^6 grains per day per plant. Pollen concentrations decreased by two thirds within 10 m of the source (a 20 x 20 m plot), while deposition at 30 m was <10% that at 1 m. Numerous dispersal models have been developed. Empirical models may describe gradients without explaining the causative processes and include the power law model and the simple exponential model, while physical models consider underlying principles of dispersal and include the gradient transfer theory (K theory), statistical or Gaussian plume models, and random walk models, reviewed by McCartney and Fitt (1985).

3.1. Terminal velocity

The principal aerodynamic property affecting the dispersal and deposition of a particle is its terminal velocity (or fall speed), which is the maximum speed to which a body falling through the air under gravity will reach. The speed of fall is prevented from increasing due to air resistance (drag). The terminal velocity of an object with a smooth surface is largely determined by its size and density. Hence opening a skydiver's parachute increases the surface area for air resistance, while the total weight remains the same, and so the terminal velocity slows considerably. This feature is used by some flowering plants, e.g. willow herb (*Chamaenerion angustifolium*), old man's beard (*Clematis vitalba*), and dandelion (*Taraxacum officinale*, Pl. 6.29), whose seeds have a fringe of hairs, a pappus, increasing drag considerably with little increase in weight and therefore decreasing terminal velocity. It is also used by some spiders, which can be blown considerable distances using long silk threads as balloons or parachutes (Schneider *et al.*, 2001). For pollens and spores, particle shape and surface texture can also affect terminal velocity. Many species of gymnosperm have winged pollen grains, for example the pollen of *Pinus* (Pl. 7.1) has two outer air-sacs, which increase buoyancy in air.

The relative effect of air resistance is greatest on objects with very small aerodynamic diameter; hence small objects reach relatively low maximum fall speeds or terminal velocities. Fall speeds of spores can generally be estimated only within ± 20% due to natural variation in spore sizes of a particular species, and hydration level as affected by the ambient relative humidity. Estimates range from 40 mm s^{-1} for large pollen grains ($\approx$ 50 µm diameter) to 0.04 mm s^{-1} for actinomycete spores ($\approx$ 1 µm diameter) (Gregory, 1976; Gregory and Henden, 1976). While terminal velocity is of high importance in very still air, normally effects of turbulence, convection or air mass movement far outweighs movements of spores at terminal velocities under the influence of gravity.

3.2. Aerodynamic diameter

The terminal velocity of a particle depends on its mass, which determines the force of gravity acting upon it, and size and shape, which together determine the drag forces opposing gravity. Many fungal spores and pollen can be approximated to spheres in shape, but some are elliptical, elongated into thin rods or fibres, or even take more complex shapes e.g. spiral, club-shaped or with radiating 'arms'. Others may be released in chains (e.g. *Cladosporium* spp. or *Blumeria graminis*) or clumps (e.g. ascospores of *Pyranopeziza brassicae* often occur in groups of four, or the rust fungus *Puccinia striiformis* may clump in humid weather into groups of seven or more spores). In order to estimate the terminal velocity and therefore the dispersal characteristics of such particles, their size and shape can be considered in terms of aerodynamic diameter, i.e. the size of a spherical object (with, for most spores and pollen, the same density as water) that would have the same terminal velocity in air. In air at 20ºC, the aerodynamic diameter d (in µm) for a spore of terminal velocity v_t is:

$$d = 18.02\sqrt{v_t}$$

when v_t is measured in cm s^{-1}. The aerodynamic diameter also affects efficiency of impaction.

Shape factors have been estimated for simple shapes such as ellipsoids and rods (Mercer 1973; Chamberlain, 1975) and can be used to estimate the terminal velocity of a non spherical spore, by dividing the terminal velocity of a spherical spore of the same volume by the shape factor (McCartney *et. al.*, 1993).

4. Deposition – sedimentation and impaction

Particles in the air descend due to gravity, eventually recrossing the laminar boundary layer and coming to rest in the still air on a solid or liquid surface (Gregory, 1973). This can be by sedimentation (passively settling onto a surface), or by impaction (the sticking of airborne particles onto a surface following an active collision) on an object's surface, e.g. a leaf or stigma of a flower because the particle's momentum may be too great to allow it to change direction and flow with much lighter air molecules around the object (McCartney and Aylor, 1987). Sedimentation can be used for trapping air particles using passive traps. Impaction is the basic principle behind many air-sampling devices such as the Andersen sampler, Hirst or Burkard spore traps, whirling arm traps (rotating arm traps or rotorods), air-filtering systems and even sticky rods. A special form of impaction, can be considered as that in which spores in the air can be removed by the action of rain. In this case, the particles impact the surface of falling rain drops, to be deposited within or on the surface of water films, depending on the particle's hydrophobicity.

5. Impact

Air-borne particles can have many effects including plant, animal and human diseases, allergies, plant pollination and colonisation of new habitats. To have an effect, particles need to have survived the airborne phase and be viable for growth, infection or pollination. In cases of allergy, however, the particle need not be viable to cause a reaction. Viability of particles in air usually decreases exponentially with time due to mortality caused by such stresses as desiccation, uv-light, starvation and extremes of temperature. The half-life of spores varies greatly from species to species according to their size, energy reserves, metabolic rate, hydration level and spore wall characteristics (e.g. pigmentation, thickness, permeability). Furthermore, having settled or impacted on a surface, there may be biochemical signalling between particle and the surface, leading to growth (i.e. germination of a spore or pollen grain) or further dormancy (and the chance of re-dispersal).

5.1. Plant disease

Plant pathogens include known species of virus, mycoplasmas, bacteria and fungi. Whilst many (virus and mycoplasmas) require an insect vector, and some others are soil-borne or water-borne, many important bacterial and fungal plant pathogens are dispersed by wind or rain-splash, and are capable of causing severe losses in susceptible crops with important economic or social consequences e.g. *Blumeria graminis* (powdery mildew of cereals, Pl. 10.30), *Puccinia striiformis* (stripe or yellow rust of wheat, Pl. 8.48-49), *Phytophthora infestans* (late blight of potato, Pl. 9.69), *Heterobasidion annosum* (conifer polypore root and butt rot, Pl. 9.43), *Mycosphaerella musicola* (Sigatoka of banana) and *Xanthomonas axonopodis* pv. *citri* (citrus canker). Aspects of the aerobiology and epidemiology of plant pathogens are investigated by plant pathologists in an effort to devise or improve disease control methods.

5.1.1. Plant disease symptom distribution

Stem and head rot of sunflowers is caused by infection by airborne ascospores of *Sclerotinia sclerotiorum* (Pl. 8.19). The number of plants infected is related to the concentration of ascospores in the air released from the apothecia on the soil (McCartney and Lacey, M., 1991a). Further work showed that the timing of the ascospore release determined the type of disease that developed (Fig. 2.3). Stem rot developed when ascospores were present before flowering and head rot when the spores were present during flowering (McCartney and Lacey, M., 1999).

5.1.2. Timing of spore release and disease control

Epidemics of phoma stem canker are initiated by airborne ascospores of *Leptosphaeria maculans* (Pl. 8.3) produced in apothecia (pseudothecia) on crop debris in the autumn. The ascospores infect leaves, leading to phoma leaf spot and eventually stem canker. In some regions, e.g. Australia, the ascospore release is well synchronised with crop emergence (a very vulnerable crop stage). By sowing the crop late, growers in some regions of Australia are able to escape the effects of canker as the spores have been released before the new crop is present. Alternatively in Europe, where the spore release is spread

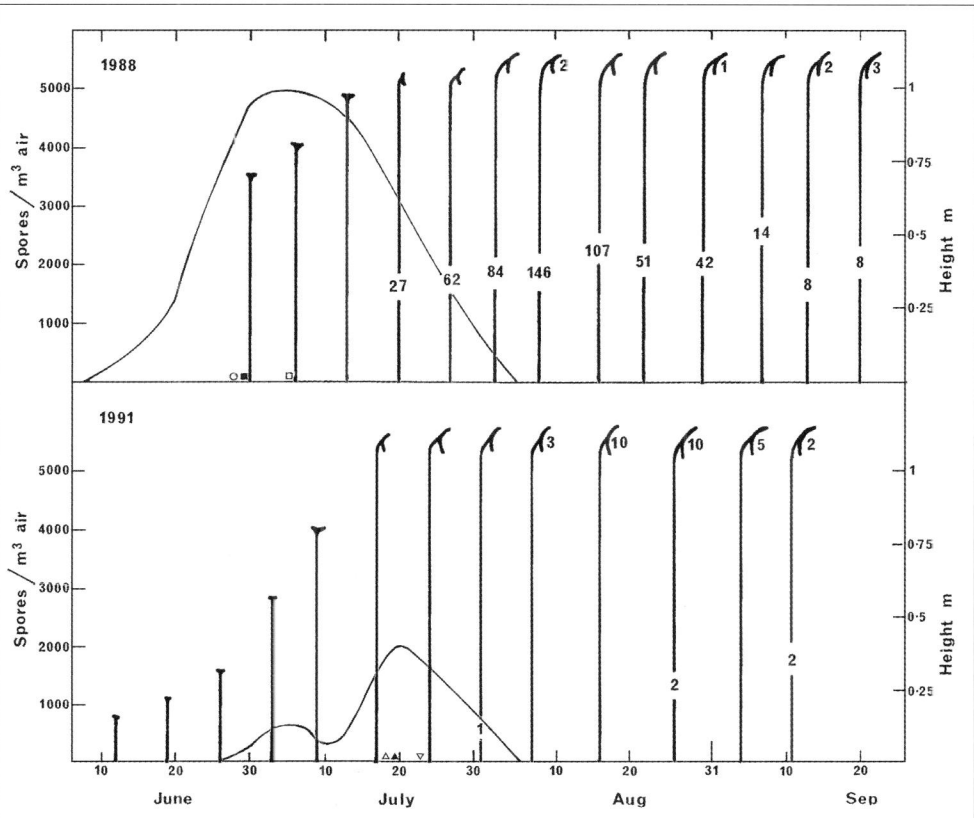

Figure 2.3
Ascospore release (line) and disease symptoms in sunflowers in 1988 and 1991. The presence of apothecia is shown by small symbols on the soil surface. Figures show numbers of new infection on the stems (at the average height) and on the seed heads.

throughout the autumn, monitoring spore release can help to target fungicide applications (West *et al.*, 2002b).

5.1.3. Disease gradients

Disease gradients can occur across fields due to spores arriving predominantly at one side of the field from a nearby inoculum source e.g. an adjacent field. Alternatively, individual disease foci can occur at random across a field when spores arrive from a distant source, producing separate patches of disease. In favourable conditions, disease severity spreads out, moving from areas of the crop with high severity to areas with low severity. Fig. 2.4 shows a near-Infra-Red image of a potato field, which shows foci of potato late blight (*Phytophthora infestans*, Pl. 8.69) as dark areas of the crop due to reduced green tissue. If favourable conditions persist, the disease patches would expand as spores from the disease foci infect the surrounding healthy areas of the crop. Due to the incubation period between infection and symptom development, an invisible zone of infection is normally already present around the visible disease foci (West *et al.*, 2003). Disease gradients are not necessarily the same as spore dispersal gradients because disease gradients are the result of many different spore production and release events over many days, each event affected by climatic factors such as wind speed and direction and incidence of rain. As a disease patch expands, the disease gradient often decreases, beco-

Figure 2.4
Aerial infra-red photograph of potato late blight (*Phytophthora infestans*), disease gradients from primary foci in a potato crop (Lacey, J. *et al.*, 1997).

ming less steep. Gregory explained that area sources of disease usually have shallower disease gradients than point sources because lateral diluting eddies would themselves contain spores rather than being spore-free (Gregory, (1976).

5.2. *Health hazards*

Many airborne fungal, actinomycete and bacterial spores are capable of causing disease in man and animals by direct infection (living tissue is invaded by the microbe), by toxicoses (ingestion of toxic metabolites of microbes), or by allergy (sensitivity to microbial proteins and polysaccharides). Respiratory allergy in man may develop immediately as in hay fever or asthma, or it can be delayed as in Farmer's Lung. Potential sources of hazardous airborne spores are many stored products including hay, straw, grain, wood chips and composts. Spore laden dust is also released into the air in many ways including distributing hay to animals, spreading out bedding and moving stored grain.

Pollen and spores are nearly always present in air but their number and type depend on the time of day, weather, season and local source. Indoors the diversity of airborne particles is usually lower than outdoors, and numbers of particles lower, unless there is a source of contamination within the building. The use of the cascade impactor and Andersen sampler together enable the different size fractions of the air spora to be monitored for both visual counts and the number of viable units. Fig. 2.5 is a very simplified diagram showing how far spores of different sizes can penetrate into the lungs and the resulting type of illness that can follow in susceptible people (Lacey, J., *et al.*, 1972).

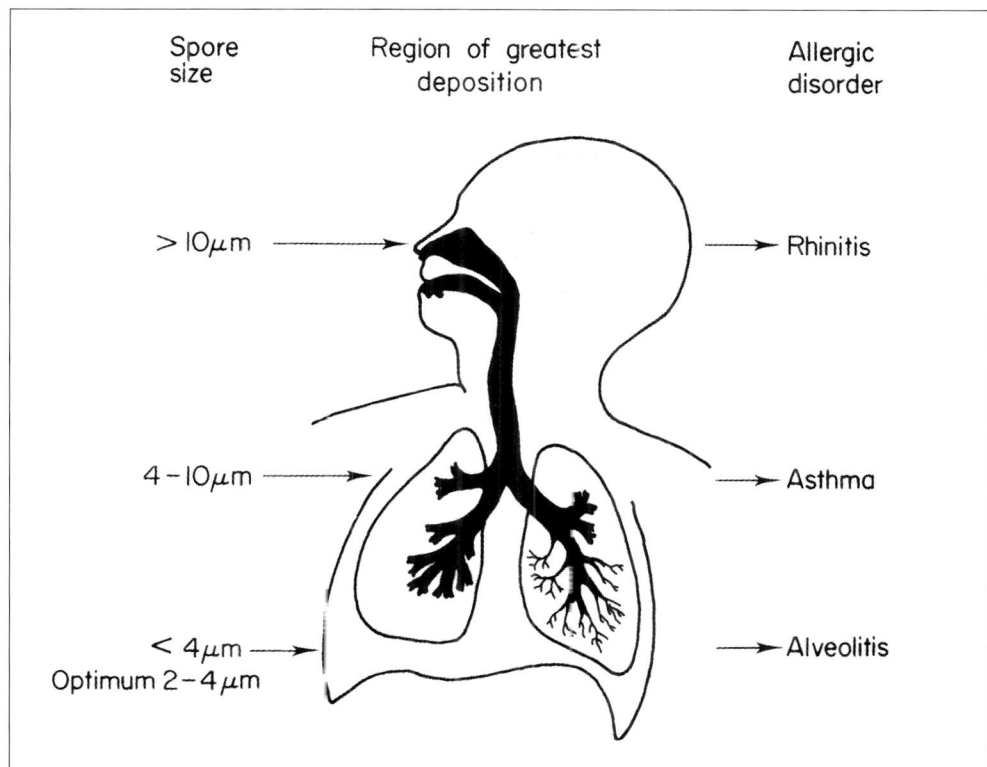

Figure 2.5
Spore size, lung penetration and type of allergic disease (Lacey, J., *et al.*, 1972, with permission from Elsevier).

5.2.1. Allergy

The increasing incidence of both pollinosis and asthma in the population at large has involved pollen and spore data being included in publications emanating from respiratory diseases, community health and medical practices (D'Amanto *et al.*, 1991; Spiewak, 1995; Emberlin, 1997; Newson *et al.*, 2000; Corden and Millington, 2001 Corden *et al.*, 2003). Pollen counts are regularly broadcast on the media and this enables sufferers have some knowledge of the presence of allergens in the air. Figure. 2.6 shows when the most common allergenic pollen is likely to be released. One of the earlier British studies investigating the relationship between pollen and spores and allergy was published by Hyde (1972). Pollen has been associated with the prevalence of allergic rhinoconjunctivitis, asthma and atopic eczema in children (Burr *et al.*, 2002). Mackay *el al.* (1992) undertook a study involving medical application of data at the Scottish Centre for Pollen Studies. The ever increasing attention to and research into the application of aerobiology to medicine is exemplified by publications involving the relationship between aerobiology and allergology (Morrow-Brown, 1994) and the airborne fungal populations in British homes and the health implications (Hunter and Lea, 1994). Between ten and twenty per cent of the world's population is considered to be city dwellers (Hunter and Lea, 1994). The changing health patterns reflect this shift from the rural environment no more so than in the increase in allergies recorded inclu-

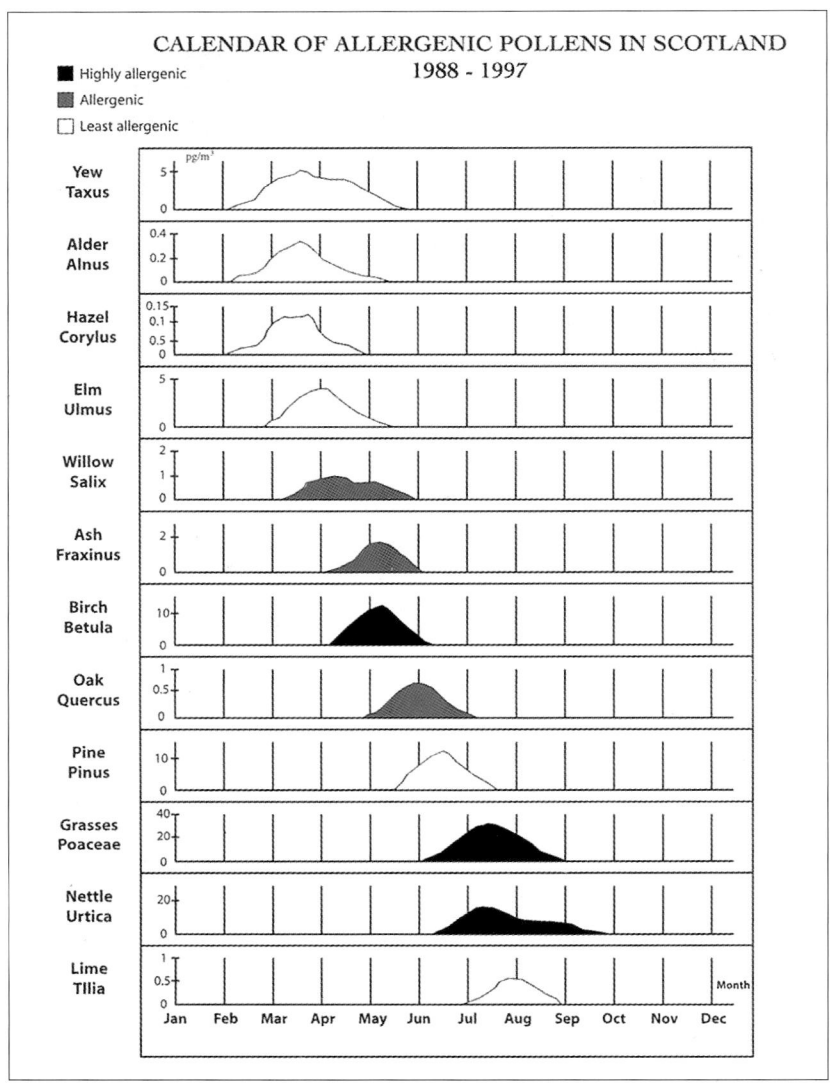

Figure 2.6
Pollen calendar showing periods when the pollen of different temperate wind-pollinated plants is likely to be in the air in Scotland. (Caulton *et al.*, 1997, with permission from The Scottish Centre for Pollen Studies, Edinburgh).

ding pollinosis, seasonal rhinitis and asthma. City dwellers spend the major part of their lives indoors working, at leisure, eating and sleeping. Public transport, schools, offices, hospitals, restaurants, libraries, community and leisure centres can all harbour pollen, fungal spores, bacteria, house dust mites, dander and other biological agents (Ranito-Lehtimaki, 1991; Reponen, 1994; Verhoeff, 1994; Nikkels *et al.*, 1996; Garrett *et al.*, 1997; Stern *et al.*, 1999 and Flannigan *et al.*, 2001).

Allergic response to allergenic pollens (Pollinosis) is not confined to humans, but also occurs in animals. Studies have been undertaken in horses (Dixen *et al.*, 1992) and dogs (Fraser et al., 2001) to identify causes of pollenosis. The methodology of these veterinary studies followed that described by Caulton (1988).

5.2.2. Late summer asthma

Many people suffer from asthma at harvest time and on dry days many spores of *Cladosporium* and *Alternaria* are in the air and can cause allergic reactions. Some asthmatic patients associated their attacks with the proximity of ripening barley. During the summer of 1972 Frankland and Gregory (1973) had a Burkard trap running in the garden of a patient in Dorset whose asthma attacks seemed to be so triggered. Large numbers of two-celled ascospores were liberated at night, similar to those seen by Last (1955), and identified as spores of *Didymella exitialis* (Pl. 8.12 and 13) produced on barley. A scientist at Rothamsted observed that his asthma usually occurred in the late summer, particularly after rain. He responded to inhalation testing with *D. exitialis* extract with an asthmatic reaction and to a skin test with an immediate reaction. Other research workers and patients were tested and it was reported in *The Lancet* that *D. exitialis* seemed to be the cause of late summer asthma (Harries *et al*, 1985). Corden and Millington (1994) confirmed that *Didymella* spores can be found in the air after rain in the summer even in an urban area.

5.2.3. Farmer's Lung

With the reduced risk of spontaneous fire in hay stacks following the widespread use of pickup balers, farmers were less cautious about making hay when it was too moist. Consequently many bales became very mouldy, increasing the problem of 'Farmer's Lung', an allergic alveolitis disease. To identify the causal agent, Gregory assembled an interdisciplinary team, including a medical team at the Institute of diseases of the Chest, Brompton Hospital, funded by the Agricultural Research Fund (Hirst, 1990). Many types of hay were examined by tumbling samples in a perforated drum at the intake end of a small wind tunnel (Fig. 2.7) and the dust caught at the other end by a cascade impactor and Andersen sampler. Visual counts from the four traces of the cascade impactor gave up to 102 million fungal spores and 1200 million actinomycete spores per g (dry weight) of hays associated with Farmer's Lung. The Andersen sampler (Andersen, 1958) allowed viable organisms to be collected dry and grown on different media at different temperatures. Predominantly thermophilic and mesophilic fungi and actinomycetes as well as bacteria were identified and counted (Gregory and Lacey, M., 1963a).

Experimental batches of hay were baled at different moisture contents and monitored for temperature, biochemical changes and mould growth (Gregory *et al*., 1963). Hay baled at 40 % moisture heated to over 60°C and contained a large flora of thermophilic fungi, particularly *Aspergillus fumigatus* (Pl. 10.16), *Absidia* spp. (Pl. 9.76-77), *Mucor pusillus* (Pl. 9.78), *Humicola lanuginose* (Pl. 10.43) and actinomycetes (Pl. 12.2-4). Extracts of the moulding hay were tested against serum from affected farmers, yielding positive reactions (Gregory *et al.*, 1964). The actinomycetes *Thermopolyspora polyspora* and *Micromonospora vulgaris* were found to be a rich source of the Farmer's Lung antigen (Pepys *et. al.*, 1963). Consequently Farmer's Lung disease was able to be registered as a prescribed disease under the National Insurance (Industrial Injuries) Act, 1964.

Spores produced from mouldy hay, shaken in a perforated drum in a wind tunnel (Fig. 2.7) at wind speeds of 0.6–4.9 m s^{-1} were sampled periodically during one hour (Gregory and Lacey, M., 1963b). The number of spores released per minute decreased rapidly from the start with two-thirds removed in the first 3 minutes. The total number released was higher with faster wind speeds. 50 million spores were released after hay was blown for 31 min at 1.2 m s^{-1}, blowing for a further 31 min at 4.9 m s^{-1} released another 55 million spores.

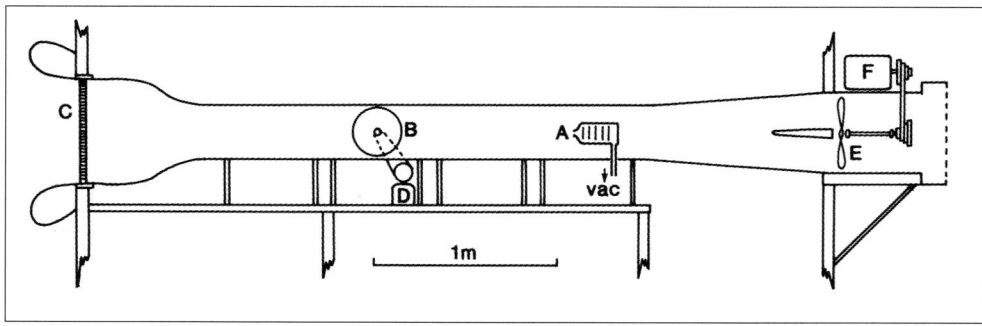

Figure 2.7
Diagram of wind tunnel showing position of collecting apparatus for studying the spore content of stored products. A, Andersen sampler or position for cascade impactor or other sampling device, B, perforated zinc drum, C, paper honeycomb, D, motor for drum, E, fan, F, motor for fan, vac, line to vacuum pump. (Lacey, J., 1990, with permission from the McGraw-Hill Companies).

Concentrations of up to 1600 million spores m^{-3} air were recorded in farm buildings while hay associated with Farmer's Lung was being shaken for animal feed (Lacey, J. and Lacey, M., 1964). Actinomycete spores were 98% of the air spora, and as they range in size from 0.5-1.3 µm in diameter, they can penetrate deeply into the lungs (Fig. 2.5).

5.2.4. *Other aerobiological hazards in the work place and home*

In addition to Farmer's Lung, there are many examples of occupational lung diseases caused by fungal and actinomycete spores (Crook and Swan, 2001; Hodgson and Flannigan, 2001). *Thermoactinomyces sacchari* was implicated in bagassosis (Lacey, J., 1971b) and *Penicillium frequentens* (Pl. 10.17) in suberosis (Ávila and Lacey, J., 1974). Further studies of the aerobiology of environments associated with occupational disease have allowed environments associated with occupational asthma and allergic alveolitis to be characterized (Lacey, J. and Crook, 1988; Lacey J. and Dutkiewicz, 1994; Crook and Swan, 2001).

An early example of research into spore or dust hazards in the work place is that for threshers during harvesting and grain storage. In the early 1970s many farm workers suffered respiratory symptoms caused by dust during harvesting of grain. Air which was being inhaled by workers on combine harvesters was sampled on farms in Lincolnshire.

The airborne dust around combine harvesters contained up to 200 million fungus spores per m³ air while drivers were exposed to up to 20 million spores per m³ air. The workers affected had an immediate hypersensitivity reaction to the spores. It was suggested that drivers could be protected by cabs ventilated with filtered air (Darke *et al.*, 1976), this is now standard practice. However, attention needs to be paid to the effectiveness of the air filters used in combine harvester cabs. Studies have shown that well fitted filters provide good protection against airborne spores, but aerosols can easily by-pass damaged or poorly maintained filters. Also, opening the cab door or window in a contaminated environment can negate the protective effect of a cab air filter within 3 minutes (Thorpe *et al.*, 1997).

The air in grain silos, sampled using a cascade sampler and an Andersen sampler while the grain was being unloaded, produced huge concentrations of bacteria, actinomycete spores and fungal spores. Many of these were viable and some were potentially pathogenic e.g. *Aspergillus fumigatus* (Pl. 10.16). It was recommended that workers should use efficient dust respirators inside silos when handling grain (Lacey, J., 1971a).

5.2.5. Compost handling and locating refuse or composting facilities

Different types of materials used for producing compost affect the type and numbers of spores released during the composting process, there can also be seasonal as well as daily changes in the number and types of spore released from composting sites. Domestic waste composts can produce high numbers of airborne bacteria (Lacey, J., *et al.*, 1992). EU legislative targets to reduce landfill disposal of waste and encourage recycling has led to a large increase in the number of green waste composting sites. However, public concern about exposure to the potentially high numbers of spores released means that the location of composting sites requires careful consideration (Lacey J, 1997). The potential for exposure to airborne spores, including *Aspergillus fumigatus*, and hazards to respiratory health associated with waste composting have been reviewed by Swan *et al.* (2002). Refuse dumps and landfill sites pose a further risk of release of potential allergens and pathogens through dry release and rain-splashed aerosols.

Mushroom compost is traditionally made from wetted straw and horse manure, which heats up during composting as many thermophilic actinomycetes grow. When moved into the growing sheds many more spores are emitted than at picking of the mushroom crop (Crook and Lacey, J., 1991).

5.2.6. Respiratory infections

In addition to allergic reactions or irritation, some airborne microbes (other than causal agents of illnesses such as colds, influenza and pneumonia) can cause respiratory system infection in humans (Campbell *et al.*, 1996; Samson *et al.*, 2001). An example, of a fungus capable of causing disease following inhalation is *Aspergillus fumigatus*, which normally grows on grain or compost. The fungus can grow saprophytically on mucus in the airways to cause bronchopulmonary aspergillosis, occasionally producing a ball of fungal growth or aspergilloma. Serious problems can occur in subjects that have suppressed immuno-systems, due to disease, immunosuppressive drugs or radiation therapy, allowing the fungus to become invasive.

Legionnaire's disease is caused by the bacterium *Legionella*, which occurs naturally in fresh water bodies such as rivers and lakes (Postgate, 1986). However, infection of the lungs, leading to serious illness, occurs if the bacterium becomes suspended in aerosol and is inhaled. While this is rare in natural systems, aerosols in buildings produced from poorly maintained cooling systems or showers pose a serious health hazard. With their growing popularity, poorly maintained spa pools are an increasing source of this respiratory pathogen. It is likely that the original outbreak of this disease in Philadelphia in 1976 was caused by infected water in the air-conditioning system resulting in an aerosol containing *Legionella* being blown into the conference hall. Other bacterial diseases that may be dispersed by air include *Bordetella pertussis* (whooping cough), *Streptococcus* species (causing sore throats, tonsillitis and pneumonia) and *Mycobacterium tuberculosis* (tuberculosis). The disease Q fever, caused by the bacterium *Coxiella burnetii*, is an example of a zoonotic disease spread from animals to humans via the aerobiological pathway. Infection can be spread from direct contact with animals or with infected bedding straw. This was thought to be the source of a recent cluster of infections in Wales (van Woerden *et al.*, 2004).

Many of the most harmful diseases, and often the most difficult to treat, are caused by viruses. Those for which inhalation is a potential route of infection include the common cold, mumps and influenza. Animal reservoirs represent a potential source and means of spread, as shown by the outbreak of the H5N1 strain of avian 'flu in the Far East (Chen *et al.*, 2004; Guan et al, 2004), while person to person spread via aerosol and droplets was an important factor in the newly emergent viral infection severe acute respiratory syndrome (SARS) (Yu *et al.*, 2004; Wang *et al.*, 2005). Foot and mouth disease virus, although not a serious human pathogen, can cause heavy economic losses through livestock infection and the disease agent is readily disseminated over long distances via the airborne route (Donaldson and Alexandersen, 2002; Gloster *et al.*, 2005).

5.2.7. Aerobiological hazards in natural environments

The impact of Pteridophyte spores inhaled in quantity constituting a heath hazard has been investigated by Siman (1999). Bracken (*Pteridium aquilinum*, Pl. 7.6) is a fern that occurs worldwide. It reproduces vegetatively but also produces a large number of spores which Evans (1987) demonstrated could be carcinogenic to mice. A ten-year study of the incidence of airborne bracken spores on an urban roof in South-east Scotland showed that they are widely present in the air (Caulton *et al.*, 1999). A Burkard trap monitored the production of spores from a small stand of bracken at Rothamsted Research from August to October in 1990 and 1991. The daily average spore content often exceeded 750 spores m^{-3}, with a maximum of 1,750 spores m^{-3}. The spore release showed a marked periodicity with most being released between 0900 and 1000 GMT (Lacey, M. and McCartney, 1994). With the large area of bracken in the UK and other countries there is potential for vast numbers of airborne spores to be present in the air and the carcinogenic properties of which cause concern to rural workers and visitors to the countryside.

6. Interpreting spore trap data

6.1. *Clumping of particles*

Conidia of *Blumeria graminis* (*Erysiphe graminis*, Pl. 10.30) in a mildewed barley crop were trapped on horizontal slides, vertical sticky cylinders and in suction traps. Spores were more often removed in clumps than singly and clumps were more efficiently deposited than single spores (McCartney and Bainbridge, 1987; McCartney, 1987). This clumping is thought to be due to the hydrophobic surfaces on the spores. Spores of *Puccinia striiformis* (yellow rust, Pl. 8.48-49) also clump together, but this is due to a mucilage surrounding them, causing the dispersal unit to be more than one spore in humid weather, while in dry conditions, individual spores are released (Sache, 2000).

In ascomycete fungi, eight ascospores are produced per ascus. In some species, groups of ascospores stick together and are captured as one particle. For example, in 1985 many hyaline cylindrical spores caught using a Burkard spore trap in a field of oilseed rape infected by *Pyrenopeziza brassicae* (light leaf spot, Pl. 8.20) were identified as ascospores rather than rain-splashed conidia because they were often in groups of four. A vertical mast with 5 whirling arm traps, and also a series of 4 traps down wind (Figs. 3.5 and 6) showed that the spores were in the air after rain and behaved as airborne spores rather than splashed spores (McCartney *et al*, 1986). In May 1986 apothecia producing ascospores were discovered on decaying leaf debris. This was the first record of the teleomorph of *Pyrenopeziza brassicae* in England, which proved important in the epidemiology of the spread of the disease (Lacey, M., *et al*, 1987).

Clumping of particles has implications for particle dispersal compared to individual spores and can affect the likely amount of disease or number of colonies formed compared to estimations based on actual spore numbers. Furthermore, different results can be obtained using different air sampling techniques due to clumping of cells e.g. an Andersen sampler measures colonies formed from particles (including clumps of cells) while other techniques such as liquid impinging or use of aerosol monitors (particles trapped on filters and subsequently suspended in liquid) may separate the clumps into individual spores resulting in higher apparent numbers of colony forming units (Crook and Lacey, J., 1988).

6.2. *Backtracking wind dispersal*

Spores of *Pithomyces chartarum* (Pl. 10.47) were first identified in Britain on spore trap slides exposed in 1958 but counted in 1960. The fungus had just been implicated in facial eczema of sheep in New Zealand and a spore had just been painted for Gregory's book. A whirling arm trap mounted in a plastic lattice-work shopping basket was carried by Gregory during a British Mycological Society fungus foray, and spores were found. Further searching up the concentration gradient over parkland enabled the fungus to be found growing on debris of *Holcus lanatus* (Gregory and Lacey, M., 1964). Backtracking of air movements is of considerable interest when considering long distance transport (see below).

6.3. Long distance dispersal

Long-distance transport of fungal spores has been demonstrated by sampling air in regions that do not produce the spores e.g. the Arctic (Meier, 1935) and over the sea (Hirst and Hurst, 1967: Hovmøller *et al.*, 2002). Hirst investigated the distribution of spores as affected by air mass movement in westerly winds. Spores were collected, using a volumetric suction trap fitted to an aircraft, during flights made from England, over the North Sea to the east. With distance, spore numbers were diluted and spores were found generally at progressively higher altitudes. It was possible to distinguish a time scale for spore dispersal due to the distribution of spore types released predominantly at night or predominantly during daylight (Fig. 2.8). Areas with a high density of spores of *Sporobolomyces* (Pl. 10.5) indicated the dispersal of night-time released spores, while pollens and *Cladosporium* spores (Pl. 11.1-2) indicated daytime discharge events (see Hirst *et al.* 1967a; Hirst, *et al.* 1967b; Hirst and Hurst 1967). Long distance transport of plant pathogens has been reviewed by Brown and Hovmøller (2002).

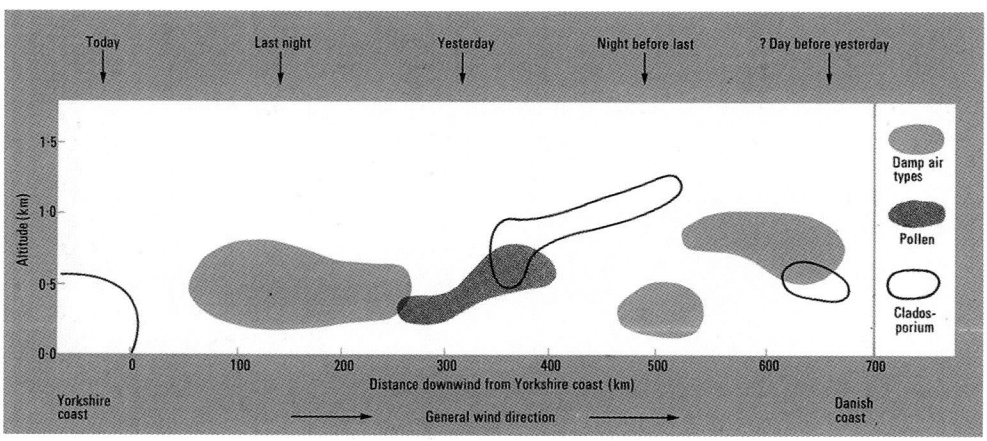

Figure 2.8 Changes in peak concentrations of pollen, *Cladosporium* spores and 'damp-air type' spores at various altitudes over the North Sea downwind of the English coast, 16th July 1964 (Gregory, 1976: from Hirst and Hurst, 1967).

Natural events that enhance long-distance transport of particles include biomass fires, which were reported by Mims and Mims (2003) to spread viable bacteria and fungal spores (*Alternaria* Pl. 11.3-6, *Cladosporium* Pl. 11.1 and 2, *Fusariella* and *Curvularia* Pl. 11.16-17) large distances e.g. over 1450 km from Yucatan to Texas. The spores were associated with coarse carbon particles collected on microscope slides (e.g. Pl. 12.29) and to eliminate contamination by local spores, a passive air sampler was flown from a kite at a Texas Gulf Coast beach. Back-trajectory analysis of the wind showed that air was travelling from Yucatan, where numerous bush fires were in progress. The authors also reported collecting spores and carbon particles at Mauna Loa Observatory, Hawaii (elevation 3400 m) on 6 July 2003, when a large fire in S. E. Asia was in progress. They speculate that convection from burning sugarcane at harvest may have helped to spread sugarcane rust (*Puccinia melanocephala*) from West Africa to the Dominican Republic in July 1978.

Turbulent weather events are thought to assist in biological particles spreading long distances. Marshall (1996) showed that cyclones moving around Antarctica were associated with a dramatic influx of airborne biological material into the South Orkney Islands from South America. The occurrence of exotic pollen trapped in moss cushions in Antarctica (Linskens *et al.*, 1993) and exotic plants found on volcanically warmed soils in Antarctica (Bargagli *et al.*, 1996; Convey *et al.*, 2000) also act as indicators of long distance propagule transfer. Similarly, turbulent weather contributed to evidence for long distance dispersal of bacteria in Sweden, where exotic *Bacillus* species were isolated from red-pigmented snow (Bovallius *et al.*, 1978). Back-trajectory analysis of wind and analysis of associated clay, fungal and pollen particulates, indicated an origin near the Black Sea, 1800 km distant, where a sandstorm had occurred 36 hours earlier.

7. Dispersal by Rain-splash and aerosol

Many plant diseases are spread during rainfall by splash and by run-off water falling to lower parts of the crop. Generally splash borne spores or bacteria are produced in mucilage which prevents dispersal by wind alone (Gregory, 1973). The mucilage dissolves on wetting to give a suspension of spores in a thin film of water on the host surface. Rain splashed fungal spores tend to be hyaline and are often filiform in shape. Diatoms are aquatic but can be blown around in the air from dried splash droplets and bursting bubbles (Khandelwal, 1992; Marshall, 1996).

Rain consists of drops of up to 5 mm in diameter falling at terminal velocity (2-9 m s^{-1}) with the larger drops (generally over 2 mm) causing the greatest amount of splash from leaves (Fitt *et al*, 1989). The amount and type of rain-splash depends on whether raindrops fall on dry surfaces or onto relatively thin or deep liquid films. The mechanism of splash was studied initially in the laboratory under simple conditions with water drops falling from known heights on to thin films of a suspension of conidia of *Fusarium solani* spread on horizontal glass slides. One incident drop 5 mm in diameter falling on a spore suspension 0.1 mm deep produced over 5200 splash droplets of which over 2000 carried one or more spores. The number of droplets deposited per unit area on a horizontal plane decreased rapidly with increasing distance from the point of impact, and in still air few droplets travelled beyond 70 cm. The larger splash droplets contained spores if either the incident drop or the surface film was a spore suspension (Gregory *et al*. 1959).

Subsequent experiments, using a rain tower at Rothamsted Research, built in order to study the interaction of rain and wind on the dispersal of plant pathogens in controlled conditions (Fitt *et al.*, 1986), showed that droplets formed from rain-splashes comprise two types: larger ballisticly splashed droplets and smaller aerosol droplets. The incorporation of inoculum into splash droplets may be considered in three stages: removal, mixing and splash droplet formation (Fig. 2.9, Fitt *et al*, 1989).

Models have been developed to describe the incorporation of pathogen spores into rain-splash droplets (Huber *et. al.*, 1996), to describe droplet dispersal (Macdonald and McCartney, 1987) and to simulate vertical spread of plant diseases in a crop canopy by

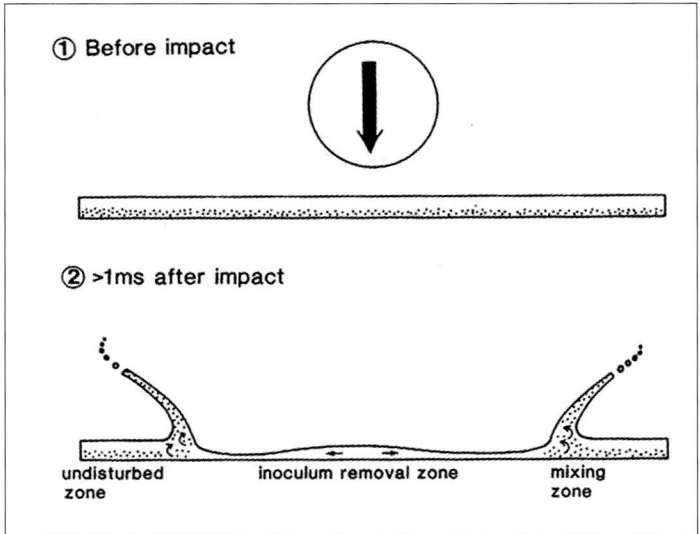

Figure 2.9
The process of inoculumm dispersal in splash droplets as raindrops strike thin films of water covering host surfsce (Fitt *et al.*, 1989).

stem extension and splash dispersal (Walklate, 1999; Walklate *et al.*, 1989; Pielaat *et al.*, 2001).

Large splash droplets (ballistic drops) tend to travel relatively short distances in wind, e.g. <16 m, but even less (<3 m) if they are subject to filtering effects of crop canopies (Stedman, 1980a; 1980b). Fitt and Bainbridge (1983) showed that spores (of *Pseudocercosporella herpotricoides*; teleomorph, *Oculimacula (Tapesia) yallundae*) are dispersed mainly in relatively large drops, but small airborne droplets may carry spores much longer distances. In fierce winds, a huge amount of windborne spray or aerosol is produced after impaction of raindrops on the ground, buildings or vegetation. Particularly strong winds combined with rain in storms or hurricanes have been important in the spread of the bacterial disease, citrus canker (*Xanthomonas axonopodis* pv. *citri*), in Florida. As a result, controversial laws currently in force to limit the spread of this disease stipulate that any citrus tree within 570 m (1900 ft) of an infection site should be destroyed. Gottwald *et al.* (2002) estimated that disease spread from a source to the nearest newly diseased tree within a 30-day period was up to 3.5 km, with infections potentially extending beyond this distance.

CHAPTER 3

Air Sampling Techniques

1. Introduction

To investigate biological particles in the air, they have to be caught by a sampling device with a design format suitable for viewing particles under a microscope or other form of analysis. Although quantitative analytical methods such as immunological and molecular techniques (real-time PCR), which can quantify target DNA accurately, are now available, this book is primarily concerned with the visual identification of airborne biological particles. As we have discussed in previous chapters, airborne particles vary greatly in size, number and type depending on time of day, weather, season and geographical location. As a result, the location of a sampling site, time and duration of operation has to be considered carefully to ensure a representative sample is taken. Furthermore, the type of sampling device used is important as each has different sampling efficiencies or suits a particular form of analysis.

Particles can either settle onto a surface by gravitational sedimentation or be impacted. Air sampling techniques described in this chapter and relying on these principles include sticky microscope slides (or other surfaces), the cascade impactor, the Andersen sampler, the whirling arm trap, the Burkard trap and the cyclone.

Much work on the sampling of the air spora has been done during the last 50 years, with methods and samplers described in the *Bioaerosols Handbook* (Cox and Wathers, 1995). Hirst (1995) in his introduction describes a **bioaerosol** 'as an aerosol comprising particles of biological origin or activity which may affect living things through infectivity, allergenicity, toxicity, pharmacological or other processes. Particle sizes may range from aerodynamic diameters of ca. 0.5 to 100 µm.' In his chapters on 'Inertial Samplers', and 'Non-inertial Samplers' Crook (1995a and b) describes many samplers that are mainly used indoors and often testing for viable microorganisms. Outdoor air sampling techniques are described by Lacey and Venette (1995). The simplest samplers collects spores passively by sedimentation onto sticky slides or agar media, or by impaction due to air currents onto thin vertical sticky surfaces. A common feature of active air samplers is the principle of impaction. Impactors use inertial impaction to capture particles on surfaces. The surfaces can be coated in a "sticky" film to retain the particles as with the cascade impactor, whirling arm trap and Burkard spore trap, or it can be a culture medium as in an Andersen sampler. Impactors are usually fairly low volume samplers e.g. the Burkard spore trap which samples at 10 l min^{-1}, but some e.g. the VWR (formerly Merk) MAS 100 Air Sampler and the Oxoid MAQS II microbiological air sampler, have sample rates as high as 100 and 120 l min^{-1} respectively. The size collection efficiency of inertial impactors depends on the speed of the airflow through the inlet, the width of the inlet and the separation between the inlet and the collecting surface. The Burkard spore trap for instance has a sample cut-off at about 1-2 µm as particles smaller than this are able to change direction and remain within the airflow. For long-term sampling there may be problems with sample loss through re-suspension or bounce-off, unless the sticky surfaces are moved past the air intake or orifice (as with the Hirst or Burkard spore traps), or the collection surfaces are continuously washed with liquid (as with wet-cyclone samplers and the SKC BioSampler. For collection onto solid surfaces the choice of collection medium is important as its efficacy depends on the ambient temperature. If the collecting surface is a clear film or glass slide coated with a sticky substance then the sample can be examined by microscopy.

Impaction samplers are available using a narrow slit to direct the airstream onto a clear glass slide coated with a sticky substance in a preloaded sterile cassette, e.g. the VersaTrap™ (produced by SKC, Dorset), can be used at flow rates of 5-30 l min^{-1} and is analysed by microscopy. Particles can impact on falling raindrops and therefore rainwater can be sampled for particles impacted on the raindrops as they fell or originating as the nucleus around which the raindrop formed.

In some recently developed formats, the sample collection surface or collecting medium is compatible with DNA analysis or immunological techniques e.g. miniature cyclone sampler (Emberlin and Baboonian, 1995), and the Burkard microtiter immunospore trapping device (MTIST) (Kennedy *et al.*, 2000)]. Even established formats such as the Burkard spore sampler can allow samples to be extracted and analysed using molecular methods (Calderon *et al.*, 2002).

2. Passive traps

This is the simplest way of collecting airborne biological particles. To trap particles by sedimentation, a microscope slide is made sticky by coating one side of it with petroleum jelly, glycerine jelly or silicone grease. The slide is placed horizontally with the sticky surface upwards. As an alternative, Petri dishes containing selective medium may be left open to sample viable propagules. These are usually exposed for 10-30 minutes but possibly longer or much shorter durations would be appropriate depending on the environment. For outdoor use, a rain-shield may be mounted above the coated-slide or Petri dish.

The technique is less efficient for small particles as due to Stoke's Law, large particles settle more quickly and consequently are deposited preferentially. The rate of settling, S, is proportional to the particles' concentration in the air ($S=Cv_d$) where the constant of proportionality, v_d, is the velocity of deposition (Chamberlain, 1975). In non-turbulent air, v_d is the same as the terminal velocity, v_t, however in turbulent air, e.g. within a crop canopy, v_d is often greater than v_t (McCartney and Fitt, 1985) and v_d is affected by particle size with the ratio of v_d / v_t for spores trapped on horizontal slides over a barley crop decreasing from about 6 for spores with v_t = 0.1 cm s^{-1}, to about 3 for spores with

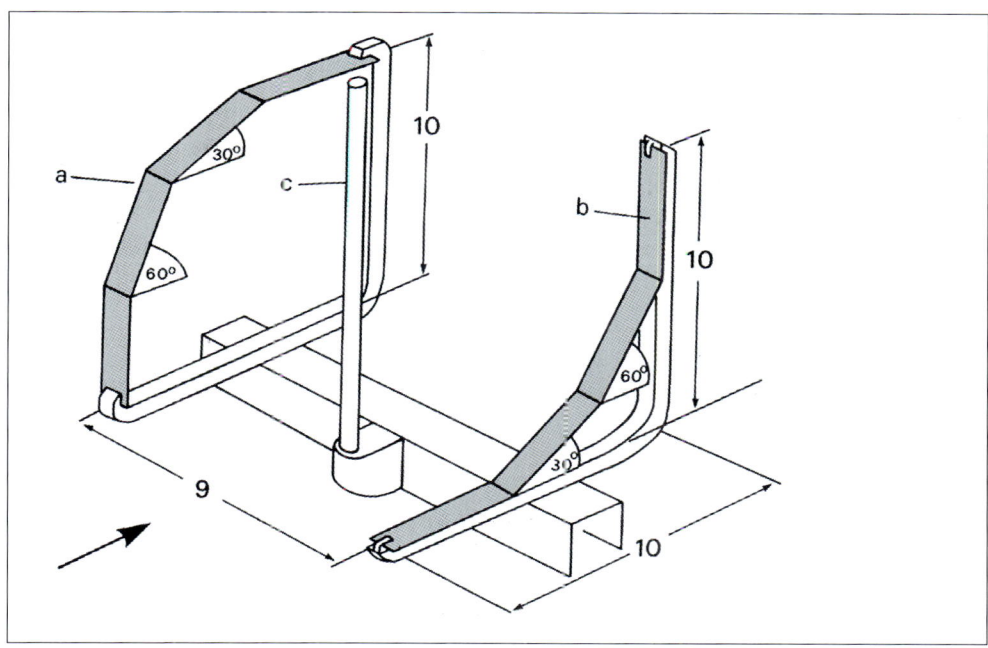

Figure 3.1 Simplified diagram of (a) convex and (b) concave shaped sticky plastic strips and (c) vertical rod used to determine the relative contributions of sedimentation and impaction of spores, units are in centimeters (McCartney and Aylor, 1987, with permission from Elsevier).

$v_t > 2$ cm s^{-1} (McCartney *et al.*, 1985). Passive trapping is a technique that is frequently used, inexpensive and easy to set up but it cannot be used for direct calculation of the concentration of particles in the air.

Passive traps can also be used to collect particles in the airstream by impaction. A thin glass rod or tube (e.g. 6 mm Ø) is coated with the same sticky substance as the glass slide technique above. The glass rod is mounted vertically so that particles impact and adhere to the rod regardless of wind direction. The glass rod samples small particles more efficiently than a slide, but again does not allow particle quantification. The rate of deposition by impaction, *I*, is proportional to the concentration of particles in the air and the wind speed, *u*, as described by I = CuE, where E is the efficiency of impaction (the rate at which particles impact on a surface divided by the rate at which they pass through the same area as the surface in the airstream). The calculation of E depends on the particle size and wind speed but the latter is complicated in field conditions by the effect of gusts of wind, which may enhance impaction rates by a factor of ten compared to I based on a constant mean wind speed (Aylor et al., 1981; McCartney and Bainbridge, 1987). Convex and concave shaped sticky plastic strips and glass rods were used to determine the relative contributions of sedimentation and impaction of spores (McCartney and Aylor, 1987).

3. Cascade impactor

The cascade impactor (May, 1945) is a four stage suction trap in which air, drawn through a narrow slit at each stage, impinges on a sticky microscope slide on which the

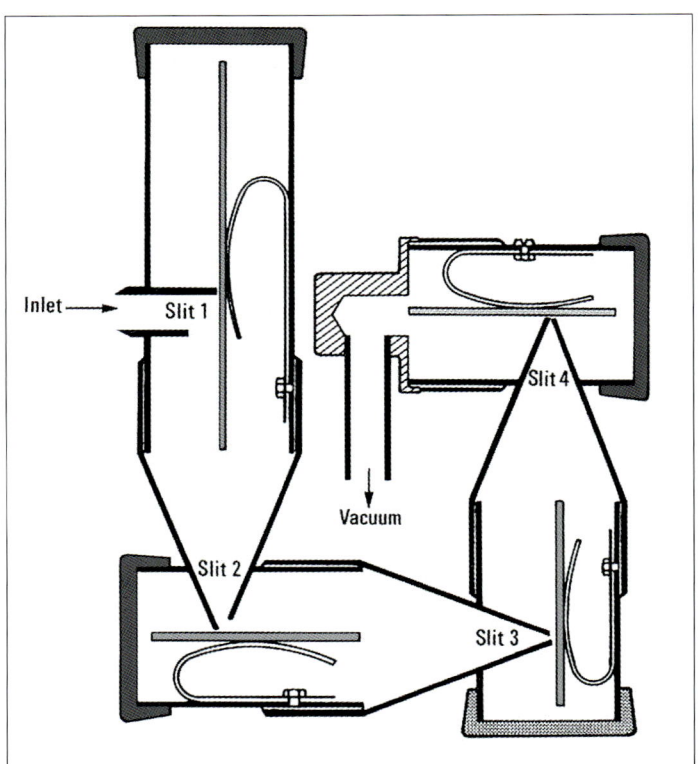

Figure 3.2 Cascade inpactor (Gregory, 1976, by permission of Oxford University Press).

airborne particles are deposited (Fig. 3.2). Each successive slit is narrower so that the air travels faster causing progressively smaller particles to be impacted.

The cascade impactor has been used extensively as a standard against which other sampling devices have been calibrated. It has also been used particularly for studying the spore content of many stored products and the dust clouds they create when disturbed.

4. Andersen sampler

This device separates and traps particles of progressively smaller sizes using a stack of perforated metal sections or stages, interspersed with open agar-filled Petri dishes and made air-tight with rubber ring-seals (Andersen, 1958; see also http://www.helios.bto.ed.ac.uk/bto/microbes/airborne.htm). The diameters of perforations in each stage reduce from the top to the bottom of the stack. Air is drawn through the sampler by an electric pump at a set flow rate. One version traps particles onto agar-filled Petri dishes to enumerate viable particles (depending on the media selected), while another version traps particles onto stainless steel or glass plates for microscopic, chemical, immunological or other form of analysis.

Although the number of colonies can be quantified with the Petri dish version, in most situations the trap is operated only for a few minutes, or less in heavily contaminated situations, before replacing the Petri dishes, otherwise they can become overloaded.

A range of different selective media can be used to sample different viable colony forming units in the air of a particular location. Incubation can also be at different temperatures. The Andersen sampler mimics the deposition of particles in the human respiratory tract (Fig 2.5), i.e. with larger particles penetrating only into the nose and trachea or first stage of the Andersen sampler and smaller particles penetrating into the lungs or to the lowest stage of the Andersen sampler. Filters can also be used instead of agar, particularly if chemical analysis of air-particles is required. Copley Scientific Ltd. produce an automated system to remove the sample from different stages of an eight-stage Andersen sampler, weighing and transferring the contents into HPLC vials and cleaning the stages in the process.

Simpler single stage samplers based on the Andersen sampler are available, often tailored to a specific market/use. The MAS 100 (VWR International) is a self-contained unit of rechargeable battery, pump and sampler, which samples into a Petri dish at 100 l min^{-1}; similarly the Oxoid MAQS II microbiological air sampler (Advances in Life Sciences) samples from 30 to 120 l min^{-1}; the Burkard portable air sampler for agar plates (Burkard Manufacturing Co. Ltd.) samples onto agar plates at 10 or 20 l min^{-1}.

5. Whirling arm trap

5.1. Description

This trap consists of a pair of vertical arms which are rotated at high speed (e.g. 3500 rpm) by a small (e.g. 12 volt dc) electric motor (Perkins, 1957). Particles are impacted on sticky-coated tape (adhesive tape) mounted on the leading edge of the rotating arms. Different versions exist but typically the arms are the uprights of a 'U'- or 'H'-shape made from a brass square-section rod with a 1.5 mm (1/16 inch) cross section. A typical example might have vertical arms about 60 mm long and ≈ 76 mm apart (Fig. 3 3).

Figure 3.3
Whirling arm trap

Tape is stuck onto the leading edge of the arms and trimmed to size. The tape strips are made sticky with a petroleum jelly/wax or silicone grease coating. After use the tape strips are mounted on microscope slides.

Alternative versions exist with detachable transparent acrylic arms that can be examined directly by microscopy. The rotating arms create an air current that enters above and below the rotating arms and is forced outwards, intercepted by the rotating arms in the process (Fig. 3.4). Efficiency of trapping is enhanced by fast rotation speed and small width of collecting surface (arms) and depends on particle size with particles over 25µm collected as efficiently as with a Burkard spore trap (Edmonds, 1972; McCartney et al., 1997). The air volume sampled by each arm can be quantified according to the equations:

$$V = \pi.D.W.L.S \text{ cm}^{-3} \text{ min}^{-1}$$

or $V = \pi.D.W.L.S. \; 10^{-3} \text{ l min}^{-1}$

or $V = \pi.D.W.L.S. \; 10^{-6} \text{ m}^{-3} \text{ min}^{-1}$

where D is the outer diameter (in cm), W the width (of collecting surface in cm), L is the length (of the collecting surface in cm) and S is the speed in rpm. The sampling rate of these traps is relatively high: the samplers used by McCartney and Lacey, M. (1991b) had a sampling rate around 157 l min^{-1}.

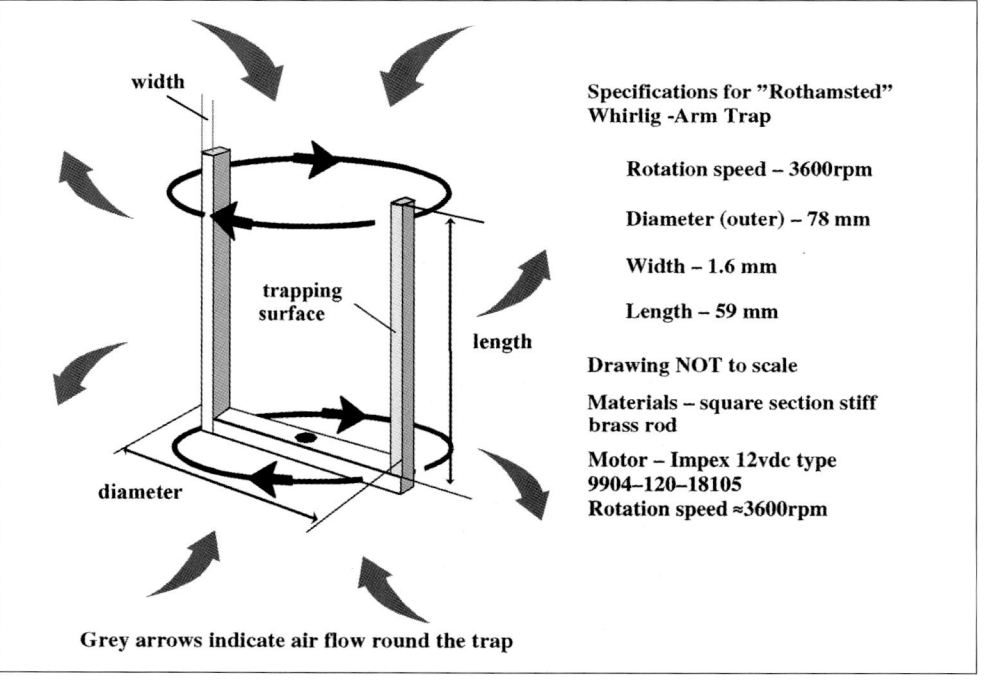

Figure 3.4 Whirling arm trap showing air movement (with permission from H. A. McCartney)

5.2. *Siting and running the trap*

The advantage of this trap is its relatively low cost, which enables several traps to be operated at one location (Lacey, M. et al., 1987), checking for gradients in particle numbers with distance from a source (Fig. 3.5) or height (Fig. 3.6). When used outdoors, shields should generally be used to protect from rain. The speed of rotation in each trap should be similar if they are operated from the same power supply. The speed of rotation of an individual rotating trap can be checked using a stroboscope. The rotating trap is set in operation, illuminated by the flashing light, of the stroboscope set to the approximate

Figure 3.5
A series of whirling arm traps down wind from an oilseed rape crop, May 1986.

Figure 3.6
Burkard spore trap and mast of whirling arm traps in an oilseed rape crop, April 1986.

rate of rotation of the motor (e.g. 3500 flashes per minute). The number of flashes should then be increased or decreased to match the number of revolutions of the trap, at which point the rotating arms appear to be stationary. The exact number of revolutions per minute for a particular motor using a particular power source is then known and can be used in calculations to quantify the volume of air sampled. If operated from a fully charged 12 volt car battery, a rotating trap can operate for several hours with little decrease in revolution speed. The traps may be mounted inside mesh cages if a large number of flying insects are present, although this may affect the sampling efficiency of each trap. Traps may be arranged with equidistant spacing or progressively further apart depending on the expected particle distribution.

Traps are usually mounted on a post or frame with the arms rotating around a vertical axis. Viewed from above, the arms rotate in a clockwise direction when the motor is correctly connected. Prepared arms are placed on the trap only immediately before operation. Arms are prepared in advance (e.g. in a clean laboratory) by placing clear adhesive tape onto the leading edge of each arm (right arm forward facing and left arm rearward facing). The tape is then trimmed back using a scalpel or razor blade so that its width and length is the same as that of the arm itself (Fig. 3.7a). The tape is then coated with a suitable sticky substance to trap airborne particles, e.g. paraffin wax and petroleum jelly (Vaseline) dissolved in hexane. This can be painted or applied to the tape surface using a plastic strip (e.g. plant label) dipped into the wax suspension and wiped along the tape surface (Fig. 3.7b).

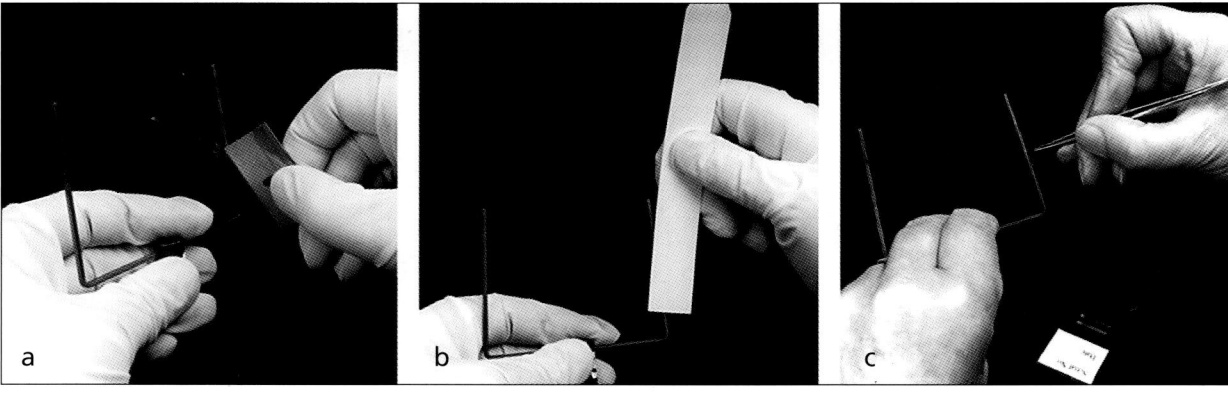

Figure 3.7
Preparing the arms, a) trimming away the excess tape, b) applying adhesive to tape strips on leading edges of arms, c) removing one of four sections of the tape strip for mounting on a slide.

Prepared arms can be stored upright in a rack inside a container (e.g. a sandwich box), to prevent contamination prior to use. A suitable rack can be made from a piece of wood drilled with a row of appropriately sized and spaced holes to hold the base of the arms with the arms upright. Each pair of arms should have a permanent serial number written on the base so that records can be kept accurately at sampling.

5.3. Using and changing the arms

At the sample site, a pair of arms is placed onto the spindle of the motor, which should be held firmly in place on a post or bracket using a clamp, jubilee clip or cable tie. The motor is connected to the battery to start rotation. After a suitable sampling period the arms can be changed by disconnecting the power to stop rotation and removing the arms from the motor spindle (handling the base of the arms). The exposed pair of arms can be stored in a plastic container (e.g. sandwich box containing a rack). The sample site can then be moved if need be, or a new pair of arms, of known serial number, replaced to begin a second period of sampling.

5.4. Mounting the tape strips

After use, the tape strips, still on the arms, are cut into four equal sections (e.g. about 1.5 cm long on a 6 cm arm), using a scalpel or razor blade. The sections are removed from the arms using forceps and placed onto a microscope slide, one arm at a time, uppermost section first (Fig. 3.7c, and Fig. 6.6). The microscope slide should be labelled with the sample number, details and date. The strips are mounted using Gelvatol or glycerine jelly, to which a stain may be added. Alternatively, the whole strip can be mounted longitudinally on the slide but this is much more difficult to handle.

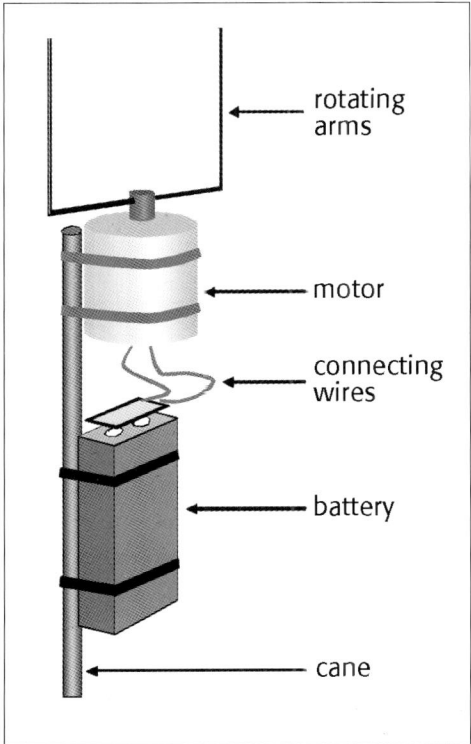

Figure 3.8
Light weight travel trap.

5.5. *Light weight travel trap*

A portable version of the whirling-arm trap can be made by mounting the motor on a small cane by tape with a small 9 volt battery attached beneath it (Fig 3.8). The device may only have sufficient power for use over short periods of time, and the speed of rotation and therefore efficiency is relatively low, but it is very effective for quickly investigating the air spora diversity in a particular environment (Plate 2.a. and b).

A mobile laboratory for collecting airborne particles would therefore comprise: prepared whirling arms, electric motor, battery, connecting wires, small cane and tape, sticky tape and adhesive, labelled microscope slides in container, cover slips, mountant, razor blade or scalpel, forceps and mounted needle.

6. Burkard trap

The Burkard trap or seven-day recording volumetric spore trap (Fig 4.1 and Fig 3.6) is based on the design of the Hirst spore trap (Fig. 1.5). It is the standard device used for pollen and spore monitoring by a number of organisations. Chapter 4 is reserved exclusively to describe its operation in detail. It comprises a chamber into which air is drawn through an intake slit by a built-in pump. The device faces into the wind by action of an attached wind vane and the air intake is protected from rain. The air-stream impacts on a slowly rotating drum, which operates by a clockwork mechanism to revolve once every seven days (2 mm per hour). The drum is covered by Melinex (a transparent plastic) tape, which is coated with a sticky wax film, on which airborne particles are trapped. The tape is replaced every week and then cut into day-length sections and mounted for microscopy. Mains or 12 volt battery-operated versions have been produced. Other variants exist, one allowing spores to be trapped on a single glass slide over a 24-hour period and one lacking a wind vane and rain cover, for indoor use.

Other similar traps such as the Lanzoni VPPS Hirst-type trap and traps working on similar principles such as the Tilak air sampler (Tilak and Kulkarni, 1970) are also available.

7. Cyclone and Miniature Cyclone samplers

In conventional wet-cyclones, a vortex is set up in a conical tube and particles are deposited on the sides of the tube by inertial forces. Water, or some other collecting liquid is used to continuously wash the cyclone walls. Wet-cyclones can be relatively high volume devices. The Biotrace Intelligent Cyclone Air Sampler (Biotrace International Plc) can sample up to 750 l min^{-1} and can be used with a luminimeter to detect ATP, indicating microbial contamination.

Dry cyclones were developed in the early 1950s (Tervet, 1950; Tervet *et al.*, 1951), comprising a glass or metal cylinder with an air intake directing the airflow tangentially

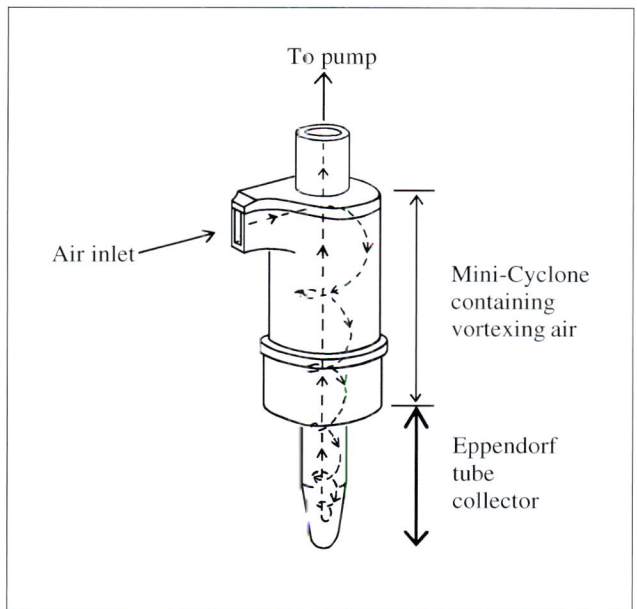

Figure 3.9 Diagramatic representation of a cyclone sampler showing approximate airflow direction within the device (with permission from H. A. McCartney).

into the upper part of the cylinder. The lower end of the cylinder is round-ended or can have a detachable sample vial into which the spores settle. The airflow leaves the cylinder via a central tube extending from the upper part of the cylinder to just below the level of the air intake.

The miniature cyclone, developed by the Burkard Manufacturing Co. Ltd., collects a dry sample in a tube. The mini-cyclone is capable of continuous sampling at 16.5 l min^{-1}, (Fig 3.9) and is highly efficient for particles as small as 1 μm. Particles are deposited into a 1.5 ml Eppendorf tube due to a change in direction of vortexing air, which causes sedimentation of particles into the tube (in a similar way to a virtual impactor). The device allows analysis of the particles by molecular or immunological techniques or by microscopy (the trapped particles would need to be transferred to a slide). It can be mounted for field use (facing the wind and protected from rain), or unmounted for indoor air sampling and sampling of deposits on surfaces (Emberlin and Baboonian, 1995; Williams *et al.*, 2001).

8. Virtual Impactors and liquid impingers

Virtual impactors use the impaction principle, but do not impact the particles onto a solid surface. Air is drawn through an "impact jet" and particles are "impacted" into a "collection chamber", moving out of the air-stream (Fig. 3.10). The collection chamber contains still air, where the particles can settle out. One of the advantages of a virtual impactor is that potentially it can sample high volumes as the flow restrictions are less than with a filter. This is the principle of the Burkard "high throughput jet" spore and particle sampler, which samples at 400 l min^{-1} (Limpert *et al.*, 1999). Rather than still

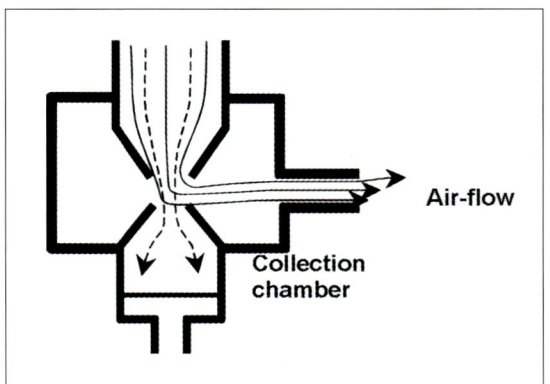

Figure 3.10 Diagram of a virtual impactor showing movement of airstream and suspended particles (with permission from H. A. McCartney).

air, the size-selective bioaerosol sampler (SSBAS), designed at the University of Turku, Finland, has a low flow rates (1.85 and 1.67 l min^{-1}) applied to filters 1 and 2 respectively in two collection chambers (Kauppinen *et al.*, 1989). The low flow rates ensures that the sample is collected onto the membrane filter, but without damaging the sample, while the total flow rate of the sampler is 18.5 l min^{-1}. The size-selective bioaerosol sampler (SSBAS) therefore simulates the human respiratory tract in both volume and fractionation and is used to collect pollen grains and smaller fractions for immunoassay (e.g. birch allergen) (Pehkonen and Rantiolehtimaki, 1994).

The BioSampler bioaerosol glass collection device, produced by SKC, uses an air inlet designed to mimic the human nose (in terms of particle size passed). The airflow from the collection chamber passes into three tangential nozzles, creating a swirling airflow inside the all-glass structure, with a flow rate of about 12.5 l min^{-1}. A liquid collection medium is swirled by the airstream inside the collector, washing particles from the inner wall of the collector. The collection liquid may be water, a liquid culture medium or light mineral oil. Samples can be processed for microscopy, culture growth, immunological, molecular or biochemical analysis. Collection efficiency is near to 100% for particles over 1 μm in diameter, decreasing to 90% for particles of 0.5 μm.

Liquid impingers are simpler devices in which air enters a container through a nozzle, the end of which is covered by a liquid. An air-outlet above the level of the liquid draws the air to a pump. Particles (and vapours) in the air are trapped in the liquid for analysis by microscopy, culturing, immunological, molecular, biochemical or optical (including flow cytometry) techniques. Day et al. (2002) used flow cytometry to detect spores of *Phytophthora infestans* (late blight of potato) using several parameters (light scatter, autofluorescence, particle width and reflectance following staining with a fluorescent brightener) to allow discrimination from other particles including unviable *P. infestans* spores.

9. Filters

Air is drawn through a filter and the particles are trapped. The size collection efficiency depends on the effective pore size of the filter used. There are a number of devices that

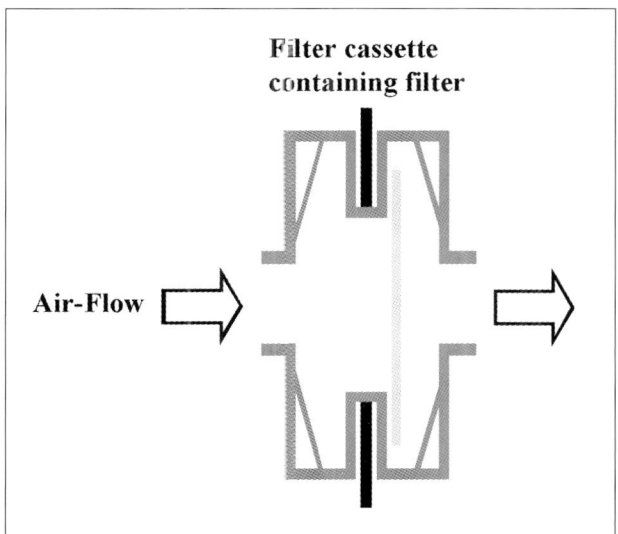

Figure 3.11 Filter air sampler in a cassette (with permission from H. A. McCartney).

use filters to collect air particles or total suspended particles (TSP). One advantage is that they are often small enough to be worn personally by people working in environments prone to dust or allergen-hazards. Samples collected on filters (especially non-fibrous ones) can be observed by microscopy either directly or by extracting the sample and suspending it in a mounting medium (although this is often difficult to do). Additionally samples can be used for DNA or chemical (toxin) analysis. Membrane filters are probably the best for this as the sample is not trapped within the filter as with a fibre filter. Sample rates are often low, unless a powerful pump is used (but this risks breaking the filter). Some filter systems can sample at reasonable flow rates, e.g. the airscan or quantifier samplers (PM10 and PM2.5) produced by SKC sample at 16.7 l min^{-1}. Large fibre filters allow the greatest flow rate. Filters have the advantage that they are easily packaged in a "cassette" form (Fig. 3.11), enabling filters to be changed in a sampling device without contaminating the collected sample. Sealed filter cassettes are produced for air sampling e.g. 25, 37 or 47 mm cassettes preloaded with Silical PVC filters are available from SKC and filter particles down to 5, 2, 0.8 or 0.2 µm. Collection size efficiency depends on the effective pore size of the filter. The flow rate depends on the density of the filter, and the power of the pump.

The Cyclone Respirable Dust Sampler (SKC) is a filter system combined with a miniature cyclone. The mini-cyclone is used to remove larger particles of dust to prevent them blocking a membrane filter, which traps the fine particles remaining in the airstream. The all-plastic device is designed to run at a flow-rate of 2.2 l min^{-1}.

The inertial spectrometer produces a deposit on a filter with particles graded according to their size. It works by the airstream being channelled through a 90° bend immediately prior to the filter. Due to inertia, the path of the larger particles deviates less than the smaller particles so they are deposited on the filter close to the bend while smaller particles travel further (Prodi *et al.*, 1979).

CHAPTER 4

Using a Burkard Trap

1. Introduction

The Burkard spore trap is used in many countries for producing the daily pollen count to help many allergy sufferers, and by scientists studying other particles in the air such as fungal spores. Running a spore trap and producing the pollen and spore counts requires considerable attention to detail. For optimal performance, careful siting of the spore trap is required along with an understanding of its operation and limitations. Trapped particles on sections of tape representing the air spora each day are usually mounted for viewing by microscope.

1.1. Safety

When working a trap using mains power, great care must be taken. The power consumption of the motor is 25 watts at 240 volts (or 50 watts at 110 volts). A good secure waterproof power supply is essential and an armoured cable may be necessary if the trap is sited some distance from the power source. Electrical cable in the U.K. is colour coded as follows; live, brown; neutral, blue; earth, yellow/green. Local Health and Safety Regulations and local electrical conventions must be observed in each country.

When handling a spore trap, as a precaution against injury (as well as damage to trapping surfaces), **always** anchor the revolving head of the trap using the rotation lock (Fig 4.1a).

2. Siting the trap

The location of the trap should be chosen according to the type of particles under investigation and the scale of interest. Traps located on the roofs of tall buildings will sample air that has been mixed thoroughly by the turbulent boundary layer. As a result, particles released from trees and the ground will be dispersed considerably so that the air sampled will represent that of a relatively large surrounding area. Conversely, due to filtering and deposition on plants, buildings and physical features, traps located at ground level are likely to reduce emphasis on sampling the background air-stream in favour of particles produced primarily from local sources. For daily pollen counts in Britain,

Figure 4.1
The Burkard seven day volumetric spore trap showing (a) principal external parts, (B) 7-day lid assembly with drum, (c) 24-hour lid assembly with slide. (BAF, 1995; with permission)

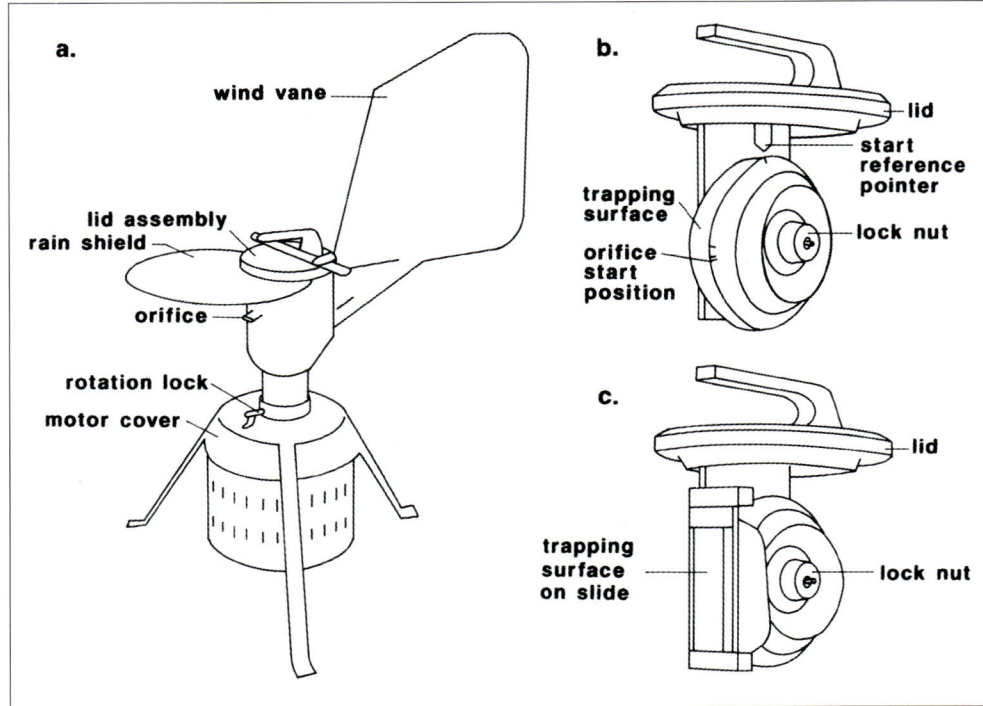

samples are collected at a network of sites using volumetric spore traps that are usually situated on flat rooftops away from obstructions and from local sources of pollen. For work in plant pathology, traps are generally placed within or just downwind of the crop to detect inoculum or to study the dispersal of plant pathogens in the field. The choice of power supply is also a consideration. When the trap is run with a mains supply within reach, an armoured cable can be used along with a waterproof and protected connection to the trap plug. Alternatively, a Burkard trap with a 12 volt motor can be run from a car battery if there is no mains supply.

3. Running the trap

The Burkard volumetric spore trap is illustrated in Fig. 4.1a. Alternative lid assemblies are available which allow continuous sampling for either 7 days (Fig. 4.1b) or 1 day (Fig. 4.1c) before the trapping surface has to be changed. The sampling rate is designed to mimic the inhalation of pollen at normal human breathing rates and is a compromise between efficient trapping and near isokinetic sampling at normal ambient wind speeds.

Great care is necessary at all stages when using a spore trap if an accurate record of airborne particles is to be produced. To avoid contamination, a clean area is necessary for the preparation and mounting of trapping surfaces. One of the best adhesives is **petro-**

leum jelly (Vaseline) which may, if necessary, be hardened with **10% paraffin wax** and is applied directly or dissolved in a solvent. Alternatively, glycerine jelly, or silicone grease in adverse weather conditions, can be used as s. It is essential that the tape or sampling area of the slide is completely covered with a thin even layer of adhesive.

At the beginning and end of each sampling period, the trace should be marked, either with a dissecting needle inserted through the inlet orifice or with a light dusting of *Lycopodium* spores from a paint brush. Such marking enables the beginning and end of the trace to be identified easily in the laboratory and enables the correct running of the clock to be checked. Also when the drum is changed, the flow rate through the trap should be checked using the **flowmeter** provided by the manufacturer, as the noise of the motor working does not necessarily mean that it is working correctly. Furthermore the clock mechanism should be wound and the intake orifice cleaned of any obstructions, such as dead insects, spiders or plant debris. It is essential, when removing the **drum**, that care is taken to ensure that nothing touches the trapping surface, and that it is placed in the carrying box as rapidly as possible to prevent contamination. Drums should be labelled at the site, with date and time of the start of the exposure and date and time of the end. It is a good idea to keep a **trapping notebook** to provide a permanent records of dates, times, flow-rates, mishaps, etc.

At the end of the season the trap should be cleaned, checked and overhauled before storing in a dry place. The manufacturer of the mains motor recommends that traps should be switched on for approximately 1-2 hours each month **while in storage** to minimise the possibility of failure at the start of the next trapping season.

3.1. *Preparing the drum*

1. Clean the **drum** thoroughly with a dry tissue.

2. The trapping tape (Melinex tape) is secured to the drum with 1 cm wide **double-sided sticky tape**. Cut a piece about 2.5 cm long and cut in half longitudinally. Place one 0.5 x 2.5 cm piece across the drum at the central mark B (Fig. 4.2a), press well down, trim to the right length, (Fig. 4.2b) and remove the protective paper.

3. Ensure that the end of the **Melinex tape** is cut square and that the edge is straight. Place it half way across the double-sided tape and press it well to secure, (Fig. 4.2c). Wind the Melinex tape tightly around the drum until it can be stuck onto the other half of the adhesive tape (Fig. 4.2d). Cut the tape off flush with the first end using a **razor blade, scalpel** or **sharp dissecting scissors**, (Fig. 4.2e), leaving no gap or overlap (Fig. 4.2f).

4. Coat the tape with **petroleum jelly** or **petroleum jelly / paraffin wax** using one of the following methods. If wax is used it should have a low melting point.

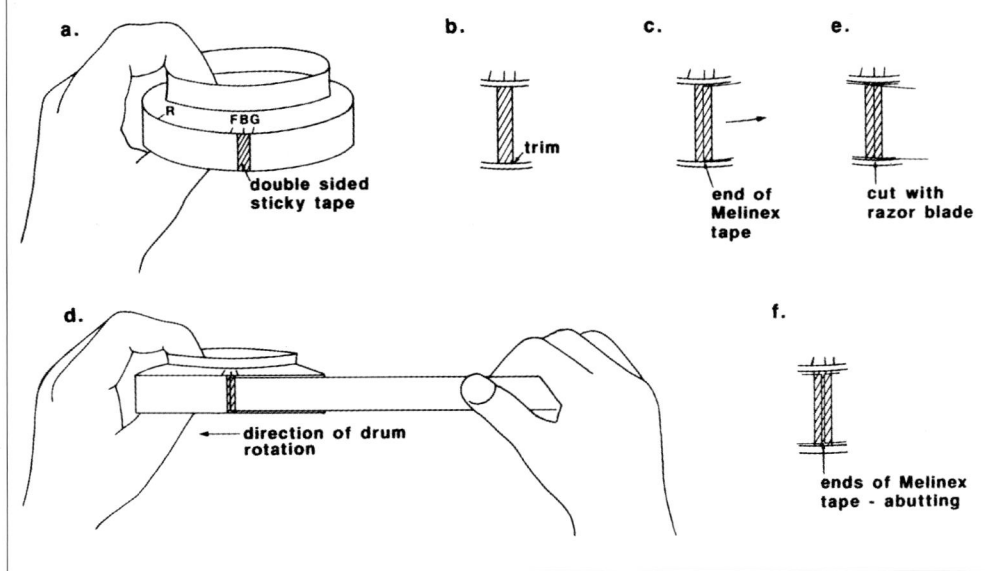

Figure 4.2 Different stages in securing Melinex tape on 7-day drum (BAF, 1995; with permission).

a) Dissolve petroleum jelly in hexane (10 g petroleum jelly / paraffin wax to 20 ml hexane) in a well ventilated area or fume cupboard wearing rubber gloves. Mount the drum on the bracket supplied by the manufacturer (Fig. 4.3a) and, while the drum is rotating, paint dissolved petroleum jelly / paraffin wax onto the tape using a paint brush (e.g. a 13-19 mm wide flat artists acrylic brush or soft hair, 19 mm wide, flat headed brush, available from BDH). Alternately, a flexible metal wallpaper stripper, cut to the exact tape width, or a plastic plant label can

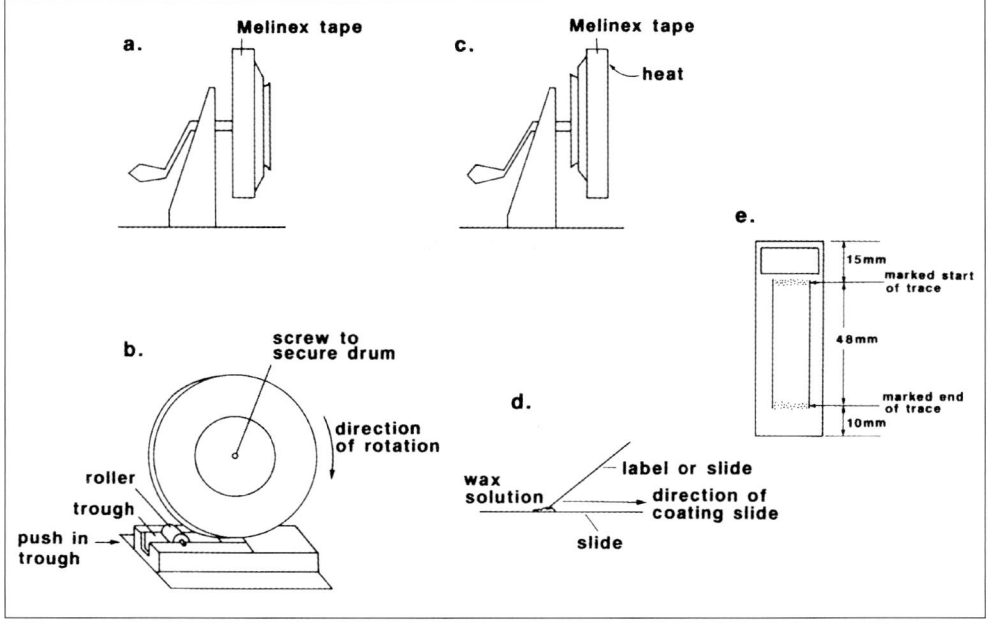

Figure 4.3 Methods of applying adhesive to the Melinex tape or glass slide surface (BAF, 1995; with permission).

be used instead of a brush. Continue to rotate the drum while the hexane evaporates to ensure even coating.

b) Dissolve petroleum jelly in hexane (10 g petroleum jelly / paraffin wax to 20 ml hexane) in a well ventilated area or fume cupboard wearing rubber gloves. If available, mount the drum on a bracket with a roller assembly (Fig. 4.3b). Fill the trough with dissolved petroleum jelly / paraffin wax and push the trough until the roller touches the drum. Then rotate the drum to enable the roller to coat the tape and, when the whole tape is evenly coated, release the trough. Continue to rotate the drum until the hexane has evaporated.

c) Mount the drum in reversed position (Fig. 4.3c) and heat it until just too hot to handle by rotating it over a flame or hot air blower. Apply hot petroleum jelly or wax mixture to the drum with a soft brush, wallpaper stripper or a plastic plant label, as above, and continue to rotate the drum until the petroleum jelly has cooled and set in a thin even layer.

If you are using a 7-day drum but only trapping for one or two days, it is only necessary to coat the beginning of the tape, i.e., about 8 cm for one day or 15 cm for two days.

Drums can be prepared in advance and stored in containers provided by the manufacturer.

3.2 Changing the drum in a trap

1. Lock the revolving head of the trap by placing the anti-swivel pin horizontally into the block of the base plate. (This is for your safety, as well as making the trap easier to work with).

2. Mark the tape through the orifice with a dissecting needle or with *Lycopodium* spores to show the end of the trace (Touch *Lycopodium* powder with finger, and snap it just in front of orifice or tap a small amount from a paint brush).

3. Depress and then swing out the locking bar on the trap, and remove the clock assembly. Care must be taken **not** to knock the clock mechanism.

4. Place the lid assembly onto a firm surface. Holding the knurled ring on the side of the aluminium drum unscrew the nut on the clock spindle, and remove the drum carefully, holding it by its central flange. **Take care not to touch the trapping surface.** Place the drum safely in a storage box and close the lid.

5. Check that the orifice is clear of obstructions. It is good practice to clean the orifice with a strip of card every week.

6. Check the airflow with the flowmeter supplied by the manufacturer and adjust the flow if it differs from 10 l min^{-1}. On mains powered traps this can usually be done by removing the white restriction orifice from the base of the sampling chamber and adjusting the screw slightly, or by cleaning the orifice and screw and then re-adjusting the screw. Replace the lid and again check the flow rate. The adjustment may need to be repeated to get a flow rate of exactly 10 l min^{-1}. On 12 volt battery powered traps, turn a knob next to the motor on the underside of the trap to adjust the rate of flow.

7. If the above procedure fails to give the correct flow rate try the following:
   - Check that the intake orifice is not obstructed.
   - Check that the 'O' ring rubber has not lifted out of position and, if necessary, push it back into its slot.
   - Check whether the 'O' ring needs replacing.
   - Ensure that the locking bar is exerting a good pressure on the lid assembly.
   - Remove the constriction orifice, replace the lid, and check whether the flow rate is greater than 10 l min^{-1}.
   - Lower the slotted motor and pump unit cover and check that the turbine is moving freely. When the spore trap is switched on, the turbine should accelerate slowly and reach a speed of 2900 rpm. At this point, considerable air movement can be felt at the end of the turbine.

8. Make sure that the clock is fully wound, turn the key anti-clockwise, **but do not over wind it.**

9. Place the new drum on the clock assembly, positioning the red mark R (Fig. 4.2a) at the arrow on the clock housing, and tighten the nut.

10. Put the clock assembly back into the trap, and secure it with the locking bar.

11. Mark the new drum, as at step 2, to show the start of the trace.

12. Hold the wind vane with one hand, unlock the revolving part of the trap, and step clear before releasing the wind vane.

3.3   Mounting the tape from the drum

1. Take the drum and cut, with a razor blade, through the double-sided sticky tape between the two ends of the Melinex tape. Lift one end of the Melinex tape (F/B), together with the sticky tape (Fig. 4.4a).

2. Transfer the tape from the drum to the cutting template, ensuring that the start of the tape (G) is placed at the left end with the 2 mm markings (Fig. 4.4b). Press down the sticky tape at F to prevent movement when cutting the Melinex tape. The cutting template has grooves every 48 mm, indicating 24-hour periods of the tape.

a) To obtain 24-hour segments that start at the same time of the day as the trap was changed, the Melinex tape should be placed on the template with the *Lycopodium* trace or needle scratch at the start (G) centrally over the first cutting grove.

b) Alternately the tape may be divided at midnight each day. The start of the trace (G) is then placed on the template at the position marking the time at which the trap was changed (Fig. 4.4c).

3. Hold the edge of the Melinex tape on both sides of the first groove on the template using forceps and cut through it, along the groove, with a sharp razor blade, scalpel or dissecting scissors (Fig. 4.4d) to remove the end of the tape and part of the incompletely formed trace (see Fig. 6.5a). Make a similar cut along the second groove.

4. **To avoid confusion always mount the trace for one day before cutting off the segment for the next.**

5. Prepare a self adhesive 11 x 22 mm label with the serial number of the sample and dates on which the trace started and ended, using a sharp pencil, and attach to the

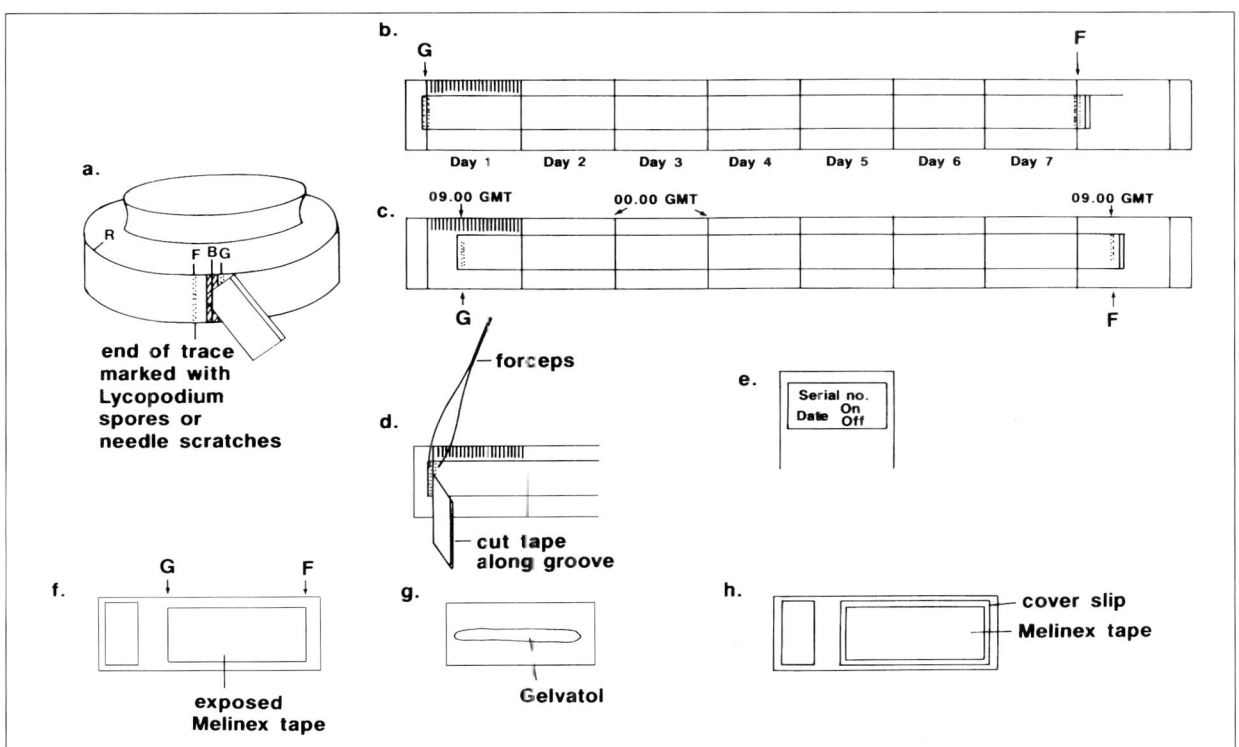

Figure 4.4
Removing Melinex tape from the drum and mounting segments on microscope slides (BAF, 1995; with permission).

USING A BURKARD TRAP

slide (Fig. 4.4e). Be consistent in the location of the label e.g. always have the label at the start (i.e. left) of the trace.

6. With a paint brush or pipette, run a film of distilled water along about 4 cm of the slide where the Melinex tape is to be placed. Hold down the cut segment of tape at one corner with fine forceps and, with the start of the tape (G) at the label end and at one end of the water film, lower it slowly onto the water (Fig. 4.4f). If any air bubbles form, lift the tape back to the bubble until it busts and then resume lowering the tape. Any remaining air bubbles that form because there is insufficient water may be displaced by lifting the corner of the tape and adding a little more water. Excess water may be blotted away with clean paper tissue, taking care not to touch the surface of the trace. The aim should be to use just sufficient water to hold the tape on the slide without air bubbles. Judging the amount becomes easier with practice.

7. With the correct amount of water the tape is just movable with forceps so that it can be positioned with the long edges parallel with the long edge of the slide, and as centrally as possible between the short edges. The exact position may depend on the microscope and the furthest point, towards the end of the slide, that can be counted satisfactorily. Some microscopes may allow the trace to be mounted almost up to the end of the slide. The best position will be found by experience. It is essential that the sides of the tape are parallel to the sides of the microscope slide so that the traverses are truly transverse during counting. If traverses are at an angle to the trace, there will be errors in determining both the time of the sample and the volume of air this represents.

8. Mark on the under surface of the slide, with a felt-tip pen, the position of any insects before they are removed with clean fine forceps. This will indicate where the trace surface has been damaged and to explain insect hairs and scales.

9. The slide needs to be protected with a cover slip before it can be examined. Usually, a mountant such as Gelvatol or glycerine jelly is used to secure the cover slip. Gelvatol can be used at room temperature but needs 24 hours to set while glycerine jelly needs to be heated before use but sets at room temperature so that it is better where counts have to be made quickly to meet press deadlines.

   a) **Gelvatol**: Run Gelvatol onto a 50 x 22 mm cover slip, using a dropping bottle, pipette or glass rod, (Fig. 4.4g). Lay the cover slip on the bench, mountant side up, and invert the slide bearing the trace and lower it gently onto the cover slip. Once the cover slip adheres to the slide, turn it over and allow the weight of the cover slip to spread the mountant evenly over the whole trace. Do not apply pressure as this could move pollen and spores. The completed slide is shown in Fig. 4.4h.

b) **Or**: place the slide on a small box to raise it a few cm above the bench. Coat a cover slip evenly with Gelvatol, using a glass rod and quickly invert it. Hold the cover slip by its short edges and gently lower it, at a shallow angle, with the lower edge further away from you, onto the slide. Allow the raised edge of the cover slip to drop slowly, possibly using two mounted needles during the final stages of descent. The weight of the cover slip will spread the mountant evenly. Do not apply pressure to the cover slip.

c) **Glycerine jelly**: Glycerine jelly is solid at room temperature and needs to be heated in its container in a water bath before use to melt it. The molten glycerine jelly can then be transferred to a cover slip as described for Gelvatol. The difference from the technique using Gelvatol is that the slide needs to be warmed to about 60°C, by passing through a Bunsen flame or by holding briefly over an electric hotplate, to allow the glycerine jelly to spread over the surface of the slide. **Beware** of overheating the slide because the jelly will boil and produce bubbles in addition to burning your fingers.

d) **Use of stains;** Staining is not essential for the identification of pollen and may obscure some important features. However staining can help to locate some types of pollen grains and hyaline fungal spores and differentiate their features. Stains suitable for pollen are Basic Fuchsin and Safranin; Trypan Blue and Cotton Blue may be used for hyaline spores. These stains are mixed with the mountant before it is applied to the slide. As it is very easy to over stain, only a drop or two at a time should be added to the mountant until there is just enough to be taken up by the target particles.

4. Using a 24-hour trap

The 24-hour trap is mainly used for studies involving short periods of sampling and particularly when information needs to be rapidly disseminated e.g. to the media.

1. Apply the petroleum jelly / wax to the slide by:

    a) **Either**: Dissolving the petroleum jelly / wax in hexane. Dip a plastic plant label (15 mm wide) or a microscope slide into the solution and then gently draw this over the labelled slide, allowing an even film of wax to form (Fig. 4.3d). It may be necessary to drain excess solution from the end of the slide by standing it briefly on a tissue. A modified flexible wallpaper stripper or soft brush can also be used to coat slides.

    b) **Or**: Gently heating the petroleum jelly / wax in a water bath to melt. Gently heat a labelled microscope slide over a flame or hot air blower until hand hot and then coat with the molten mixture using a brush.

2. Slides should be prepared with adhesive covering about 18 mm of their width, placed centrally. Any excess should be scraped away. Prepared slides should be stored in a covered glass staining dish or any other suitable container.

3. To change slides on a 24-hour trap, use the same technique as for 7-day traps, except that the new slide replaces the old one and the slide holder has to be pushed back down to its starting position. The position of the trace on the slide is shown on Fig. 4.3e.

4. Exposed slides are mounted directly, using Gelvatol or glycerine jelly in the same way as described for individual traces from the 7-day trap.

The prepared slides should be stored in a suitable slide box and kept for counting and reference.

CHAPTER 5

Using a microscope

1 Introduction

A microscope is a complex precision instrument used by many biologists. It is essential for those who use a microscope for identifying and counting spores and pollen on slides to be familiar with their microscope and its operation. The best lighting and contrast is needed to enable critical features of pollen and spores necessary for their identification to be seen.

2 Microscope Structure

2.1 Introduction

The lower limit of perception of the human eye is approximately one tenth of a millimetre (0.1 mm) in diameter. A microscope is necessary if details of objects smaller than this are to be seen and studied. Most modern microscopes are compound microscopes, in which a system of lenses is used to compound, or progressively increase, the magnification of the object being studied (Fig 5.1). Additionally, substage structures can be used to alter the width of the field of vision and light intensity. The progressive increase in magnification is achieved by combining, in a rotating series, increasingly powerful lenses up to the limit of resolution of the microscope, which is the degree of clarity at maximum magnification possible with transmitted light. The compound microscope consists of three parts held in place by a stand comprising a stout and often heavy horizontal base supporting a vertical or inclined limb. The principal parts are:

- The eyepiece(s) and objectives, the system of lenses which magnify the specimen.
- The stage, which allows movement of the specimen under the objective.
- The subtage condenser and lighting systems.

2.2 The eyepiece system

Microscopes are either monocular (with one eyepiece lens) or binocular (with two) Both types of microscope are in common use but most recent (and more expensive) models are binocular. With these, the two eyepieces can be moved apart or together to

Figure 5.1 A typical binocular microscope showing the principal parts (BAF, 1995, with permission)

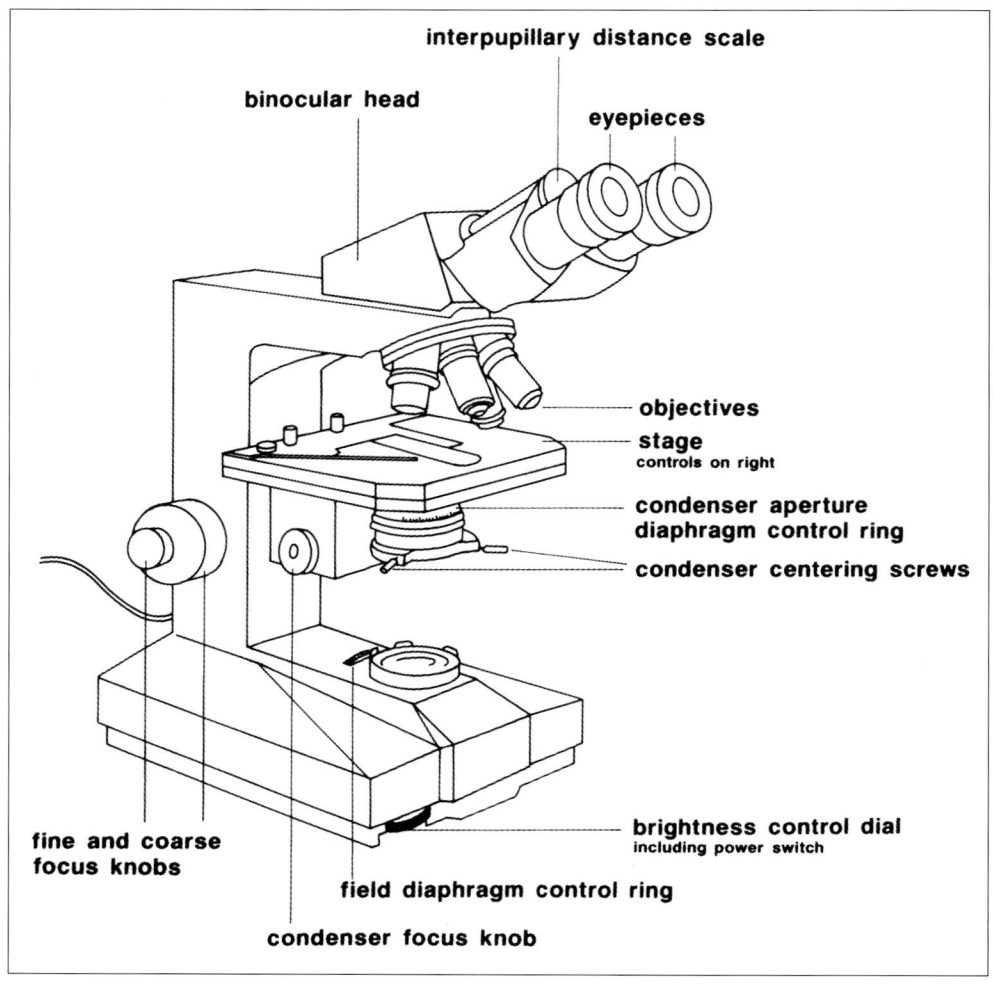

match the separation of the user's eyes. A small numbered scale is often placed between the eyepieces, on the head, to allow the precise separation to be noted for further use. Eyepieces can be of different magnification but most commonly x10 is used.

As most people's eyesight is different from one eye to the other (sometimes even when corrected with spectacles or contact lenses), it is necessary to focus the eyepieces to suit the user. Normally one of the eyepieces is fixed and the other has a focussing ring. A slide is viewed and the image brought to sharp focus to the eye above the fixed eyepiece (the other eye closed). Then the other eye is used and the image is brought to a sharp focus not by using the microscope's main focussing knobs but by adjusting the eyepiece focussing ring. The eyepieces are now focussed for the user and the position of the focussing ring may be noted to avoid repeating this if the microscope is used by different people.

2.3 The objectives

Most compound microscopes carry four to six objectives. Commonly used objectives are x4, x10, x20, x40, x60 and x100 (oil immersion). When viewing an objective under the microscope with a x10 eyepiece, these objectives will achieve magnifications, of 10 x 4 = x40, 10 x 10 = x100, 10 x 40 = x400, 10 x 60 = x600. and 10 x 100 = x1000, respectively. Each objective is normally marked with its magnifying power on the side and is often colour coded in a thin band around the lower end of the objective e.g. yellow, green, blue and red. The objectives are screwed into a swivel nose piece, which can be rotated clockwise or anticlockwise to bring any given objective into line with the specimen being studied. Most modern microscopes are parfocal, so that whichever objective is swung into place above the specimen by rotation of the nosepiece, the specimen will be more or less in focus. Fine focusing to improve the sharpness of the image is achieved with the minimum turning of the fine adjustment knob. Phase contrast objectives can also be used but need extra lighting.

2.4 The stage

This is a horizontal square plate with a central hole for the passage of light. At its simplest the stage is fitted only with a pair of spring clips to secure the slide. Such a simple arrangement requires the slide to be moved manually when viewing different parts of a specimen. This is very difficult to achieve accurately under higher magnification and to overcome this problem most compound microscopes have a mechanical stage screwed to the upper surface of the fixed stage plate. The mechanical stage consists of a horizontal slide grip with levered side arms to allow for the insertion of the slide into the movable carriage. The carriage bearing the slide can be moved sideways and forwards and backwards by two knobs set horizontally and vertically at one side of the stage. Left-handedness may require the choice of a mechanical stage with coarse and fine adjustments on both sides of the microscope stand. Two vernier scales, one to the front or rear and the other to one side of the stage enable the position of an object to be defined (Fig. 5.2). This is useful when it is desired to define the position of a particular feature on a

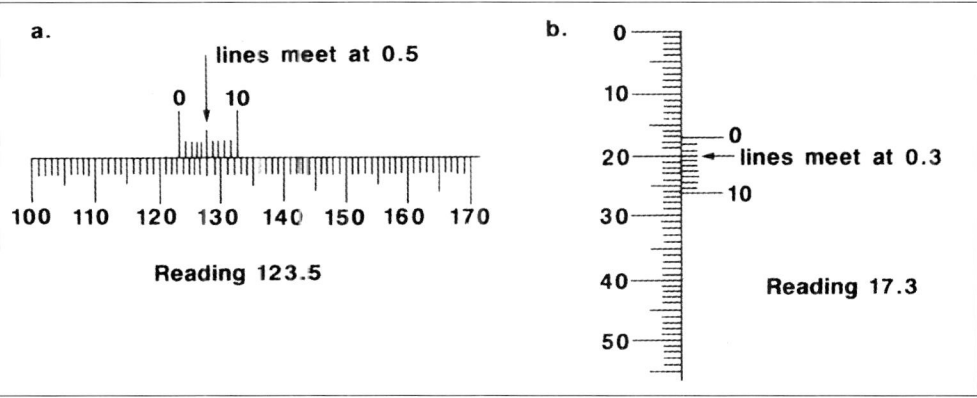

Figure 5.2
Vernier scales, a) horizontal scale, b) vertical scale (BAF, 1995, with permission)

USING A MICROSCOPE

specimen, or a particular pollen grain or fungal spore so that it can be found again. Its use is also essential when measuring the position of the traverses used when counting pollen and spores.

2.5 The substage lighting system

Centrally placed beneath the stage is a conical lens set in a vertically movable carriage. This lens, called the condenser, transmits vertically and concentrates the beam of light from the light source below, focusing the beam to run parallel through the specimen and onto the objective lens. The walls of the cone are blackened, and only the flattened apex and basal lens are clear. The condenser can be raised or lowered to alter the concentration of the beam of light passing through. A knob near the side of the condenser controls this movement. The field of view seen through the lens system can be narrowed to almost a pinhead by means of a diaphragm which is operated by a finger lever or control ring above the condenser.

Microscopes with built-in light sources have a flex with a 13-amp plug (in the UK) attached. Sometimes the flex may connect to the base of the microscope with a plug and socket.

2.6 Illumination

Both natural (daylight) and artificial lighting have been used with microscopes. Older, simpler microscopes are fitted with a movable plano-concave mirror that is used to direct the daylight or light from a bench lamp into the microscope. The concave mirror is used with daylight to concentrate diffuse light beams and bend them towards the centre of the stage through the condenser. Light from an artificial source is already concentrated so that the plane mirror can be used to direct the light beam into the condenser. However most modern compound microscopes have a lighting system built into the base of the stand and light is directed through a ground glass diffuser. The bulb should be centred when replaced or when the microscope is moved. Light intensity can be adjusted with a control set either in the base of the instrument or in a separate transformer where this is provided. A movable ring above the light source takes circular glass discs to filter out certain wavelengths of light. The colour filters can be inserted and removed by hand, the ring being swung outwards by means of a finger lever. This filter ring lies immediately beneath the diaphragm. Filters are generally not essential for pollen and spore identification but a blue coloured filter provides a whiter light for viewing and photography.

2.7 The stand

All the above systems are arranged in sequence on a strong metal support with a heavy base to ensure a low centre of gravity to make the microscope stable. The vertical part of the support carries the eyepiece head with its lenses and the stage with its substage attachments. The eyepieces are often inclined at 45° to allow comfortable viewing. The

Figure 5.3
Examples of eyepieces: (a) eyepiece scale graticule; (b, c) different types of squared graticules (BAF, 1995, with permission).

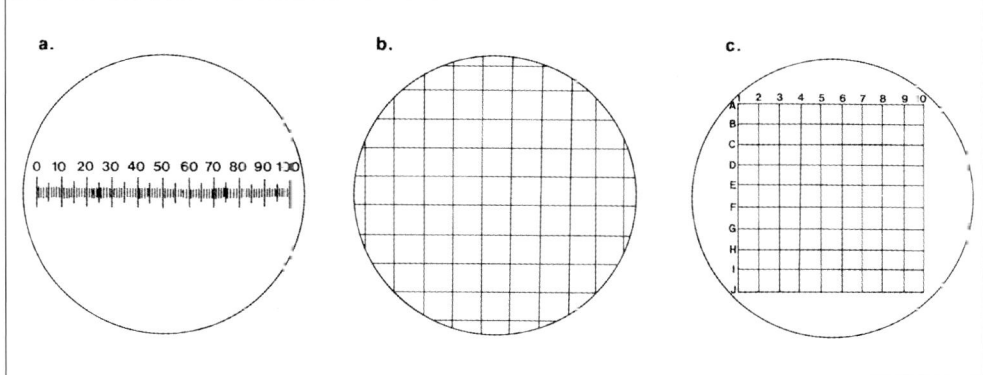

lens system in the swivel nosepiece can be moved up and down for coarse and fine focusing by means of large and small knobs set in the sides of the vertical limb, so that either hand can be used for focusing. The larger wheel/knob provides the **coarse adjustment** and the smaller wheel/knob the **fine adjustment**. Usually when the adjusters are moved in a clockwise direction (i.e. away from the operator) the lens system is lowered or stage raised. The reverse occurs when the adjusters are rotated in an anti-clockwise direction towards the operator.

2.8 *Eyepiece graticules*

An eyepiece graticule consists of small glass disc, slightly smaller than the diameter of the eyepiece tube, which rests on a ring shelf in the lower half of the eyepiece. One surface of the graticule is etched with a scale. The scale graticule is divided centrally into a hundred equal parts, numbered 0-100 (Fig. 5.3a), other graticules consist of a grid of squares of equal area (Fig. 5.3b), or have squares, 1 mm x 1 mm, which are labelled 1–10 on one side, and A–J along the adjacent side (Fig. 5.3c).

2.9 *Stage micrometer*

The stage micrometer consists of a scale etched on the top of a microscope slide and protected by a sealed cover slip (Fig. 5.4). This scale consists of a horizontal line of divisions, 1 mm in length, divided first into ten equal parts (0.1 mm) marked by long lines, with each division further divided into ten equal parts (0.01 mm) marked by short lines.

A stage micrometer is a precision scale and is very expensive to manufacture and must therefore be handled with great care to avoid damage. To avoid confusion with other permanent slide preparations, the whole slide, apart from the cover slip, is often coloured black or covered with coloured adhesive paper. The stage micrometer is usually supplied in a small plastic foam padded container to protect it from damage. The slide should always be kept in its box when not in use and stored carefully.

Figure 5.4
Stage micrometer showing appearance of the slide and the scale, as seen through the microscope (BAF, 1995, with permission).

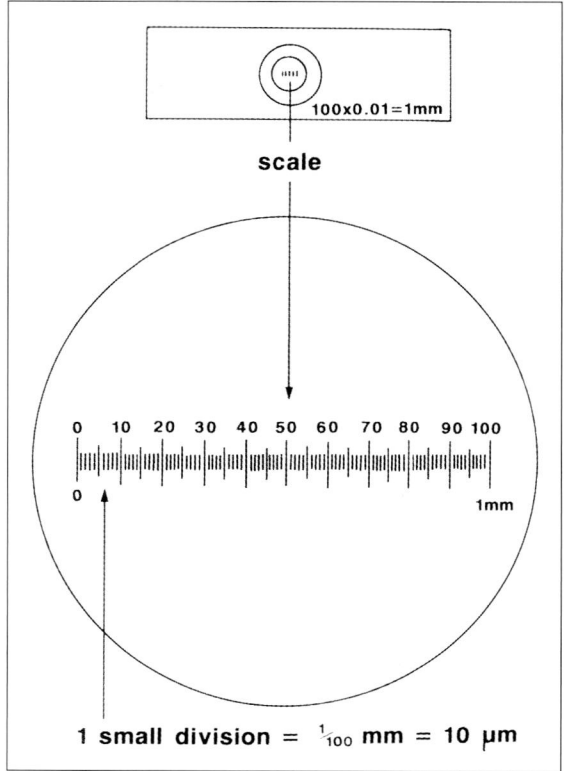

Table 1 Understanding the measurements of the divisions of a stage micrometer.

No: divisions	Length in mm	No: µm
100	1	1000
10	0.1	100
1	0.01	10
0.1	0.001	1

2.10 Protection

The microscope is usually supplied with a strong plastic cover, often loosely moulded to fit over the microscope. The cover should always be used when the microscope is not in use to protect the lenses and other parts from grit or dust raised by movement in the laboratory. Paper tissues should be used to dust the stage, and lens tissues (never other tissues) to clean eyepiece and objective lenses. It is better to remove spectacles but if they are necessary, plastic cups should be attached to the eyepieces to avoid scratching the lenses.

2.11 Camera and drawing tube

A camera can be attached to the microscope by a trinocular head. Photographs can then be taken of the objects being viewed (Plates 1-3). A drawing tube can also be attached enabling accurate representations to be made as the drawing (or painting, Plates 4-12) can be seen superimposed on the object being viewed down the microscope (Lacey, M., 1997).

3 Using a microscope

3.1 Operation

A microscope will function well if it is operated with care at all times. It is heavy but can be transported, if necessary, with one hand firmly gripping the support arm and the other supporting the base. However, microscopes should not be moved too often from the bench where they are regularly used. When not in use, the microscope's light switch should be turned off, electric power should be switched off at the socket, the plug removed and the plastic cover put in place.

When first using the microscope it is strongly advisable to follow the same set of procedures each time so that they will eventually become habitual.

1. Remove dust cover.

2. Check stage and lenses are clean and dust free (a clean objective lens is often more important than its quality – lenses should be cleaned with lens tissue only, including the condenser lenses and any in the field diaphragm).

3. Plug in and switch on the power.

4. Ensure a low power objective (e.g. x4 or x10) is in position.

5. With the coarse focus knob, raise the objective about 2.5 cm above the stage to leave adequate room to insert the slide into the holder on the mechanical stage without scratching the objective lens.

6. Switch on the microscope light.

7. Place a slide on the mechanical stage. Never insert or remove a slide with an objective close to the slide as the objective lens can be scratched. The slide used should be clean, with the cover slip free of old immersion oil or mountant.

8. If necessary adjust the distance between the eyepieces on the binocular head by moving the eyepieces apart then slowly together until the two images exactly coincide.

9. Ensure you are seated comfortably close to the bench, back straight to avoid tension, the stool/seat sufficiently high to allow ease of viewing and working. A foot rest can be beneficial.

10. Adjust the light intensity for comfort and to avoid eye strain when using low power objectives at the start.

11. Lower a low power lens to bring the specimen into near focus using the coarse adjustment while viewing through the eyepiece(s) then sharpen the image using the fine adjustment.

12. Select the power of objective lens to be used and adjust the fine focus to give a sharp image of the specimen.

13. Set the condenser for the objective lens:
    a) Close field diaphragm to give a small point of light.
    b) Adjust the condenser position using the condenser focus knob (Fig 5.1) so that a sharp image of the field diaphragm is superimposed on the sharp image of the specimen (usually the condenser is near the bottom of the slide at this stage).
    c) Centre the image of the field diaphragm using the condenser centring screws.
    d) Open the field diaphragm until no more than the field of view is illuminated.
    e) The condenser diaphragm is then adjusted to produce a circle of light between 70% and 90% of its maximum when viewed with an eyepiece of the microscope removed (this corresponds to the back focal plane of the objective lens).
    f) Replace the eyepiece.

3.2 *Focusing under high power*

Focusing with high power requires more careful adjustment than with low power.

Turn the swivel nosepiece so that a higher-powered objective, e.g. x40, comes into position. Observing continually through the eyepiece(s), slowly bring the specimen into sharp focus. Where the condenser has different lenses for different objectives, check that the correct condenser lens is in place for the high power magnification. To ensure that neither the objective lens nor the specimen is damaged when fine focusing, especially under high power, always assume the image to be over-focused. Turn the fine adjustment anti-clockwise a fraction while observing the specimen through the eyepiece(s) and if the image gradually fades, the fine adjustment can then safely be turned clockwise to bring the specimen into sharp focus. Modern objectives are often 'spring loaded' so that the lens compresses upwards if contact is made with the cover slip. Careful focusing should avoid such direct contact but, under high power, the focal end point comes rapidly, so care is essential. Adjust the condenser position and condenser aperture diaphragm to optimise illumination further. At high powers, e.g. x100, the depth of field is small so that parts above or below the centre of a relatively large specimen or relatively thick section will be out of focus and hence will decrease resolution. So high powers are only appropriate for observing detail of relatively small or thin objects.

3.3 Using oil immersion

The highest power objective typically gives a total magnification of x1000. This is close to the limit of resolution that is obtainable under the light microscope. The higher the magnification the smaller the diameter of the lens, the less light can be transmitted and greater light intensity is required. When light waves pass from the light source to the eye they pass successively through alternate layers of glass and air, each time this happens less light is available to reach the eye. The loss of light under very high power lenses through the thin separating layer of air (gas) and between the objective (glass) and the cover slip (glass) is at its greatest. To remedy this the lens is immersed in a drop of oil. Immersion oil has the same power to bend light waves passing through it as glass (i.e. the refractive index is similar). After using an oil immersion lens, it should be cleaned immediately and the slide should also be cleaned before rotating the nosepiece so that other lenses are not contaminated.

When using an oil immersion objective (x100) the following procedure should be adopted (assuming the pollen grain or spore to be examined is centrally placed and will come into view when the oil immersion objective is lowered into place).

1. Focus under low power, centre the object to be examined, then move the low power objective out of the way. Using the pipette fitted-cap place a drop of immersion oil on the cover slip and replace the cap immediately. Move the x100 objective into position.

2. Lower your head to bring the eyes level with the stage.

3. Using the coarse adjustment, slowly lower the objective until it makes contact with the oil drop. Stop at this point.

4. Raise the head and, looking through the eyepiece(s), very slowly lower the objective to a sharp focus, remembering that the end-point comes rapidly.

5. Since the light intensity had been adjusted for use with a lower power objective, it should now be increased to improve clarity and comfort. The condenser may also be raised or lowered to give maximum clarity.

3.4 Practice

A little time at the beginning of each session adjusting the optics and illumination to suit your eyes, particularly if you are using a different microscope, is worthwhile. Light intensity should be adjusted according to magnification. Generally, the greater the magnification the higher the light intensity required. Working at a microscope for long periods of time is not recommended, even when you are used to microscopy. It is useful to have some pictures on the wall behind your microscope so that there is something with a distant view to look at to rest your eyes regularly. Frequent short breaks away from the bench to rest the eyes and exercise the body muscles is also advisable. When leaving the microscope for such rest periods, always rack up the objectives to avoid possible damage to lens and slide in your absence.

3.5 Using vernier scales

The microscope slide is placed on the stage with the clip holding it steady. For Burkard spore trap slides, time of day is read on the horizontal scale (Fig.5.2a) and the position across the slide on the vertical scale (Fig 5.2b).

1. To define the position of the stage, read the value on the large scale immediately before the '0' line of the small scale. This gives the whole 'numbers' of the position. (e.g. 123 on the figure).

2. To determine the decimal value of the exact position of the stage, determine which of the small scale lines exactly opposite a line on the large scale. This value on the small scale is the decimal value of the interval between the whole number position on the large scale and the '0' position on the small scale. In one example (Fig. 5.2a) this is 0.5 making the exact position 123.5 (123 + 0.5) and in the other (Fig. 5.2b), 17.3 (17 + 0.3). The position on the slide is written as 123.5 / 17.3.

3.6 Ending operations

The procedure when microscopy is completed is almost the reverse of starting.

1. With the coarse adjustment, rack up the lens system.

2. Turn the nosepiece until the lowest power objective is in place.

3. Remove the slide from the stage.

4. Turn down the microscope's illumination control to zero, then switch off.

5. Switch off the power at the socket and remove the plug.

6. Clean the stage with tissue to remove any oil or liquid that may be present.

7. If oil immersion has been used, clean the objective lens with lens tissue – **do not use solvents or ordinary tissue**.

8. If your slide is permanently sealed, carefully remove the oil using a tissue. A drop of 70% alcohol can be used to polish the cover slip.

9. Lower the lowest power objective to within 2 cm of the stage.

10. Replace the dust cover.

> **The microscope is a sophisticated and precise instrument, but it is, nevertheless, a tool. Like all tools, its proper handling requires practice. What it can do, and above all, how well it can perform depends on the operator. Get to know the instrument, what it can do and how you can achieve the maximum efficiency of use with eyepieces, objectives, focusing adjustment, illumination, etc. Learn how to make it do what you want, in the way that you want. This can only be achieved with practice.**

4 Measuring and calibration

To determine the size of an object under the microscope, it must be measured against a defined scale, known as an **eyepiece graticule scale** (Fig. 5.3a). It is essential to know precisely the values of the graticule scale divisions used for each magnification. Knowing the magnification provided by the microscope's lens is insufficient to determine the size of the object viewed, therefore calibration of the divisions of the graticule scale is needed.

4.1 *General remarks on calibration*

Calibration of an eyepiece graticule scale, often left permanently in an eyepiece, is achieved by measuring its divisions against a slide micrometer, (Fig. 5.4) placed temporarily on the stage. Table 1 shows the different lengths in mm and µm of different numbers of small divisions of the stage micrometer.

> **Each microscope has to be calibrated individually for each magnification, as differences can occur even between microscopes of the same make and model. These differences could be vital at high magnifications.**

Calibration begins by unscrewing the upper lens of the eyepiece and inserting a graticule. To settle the disc horizontally on its shelf, gently tap or shake the eyepiece tube. Replace the upper lens and view though the microscope to check that the scale reads

Figure 5.5 Calibrating an eyepiece scale graticule against a stage micrometer with different objectives (BAF, 1995, with permission).

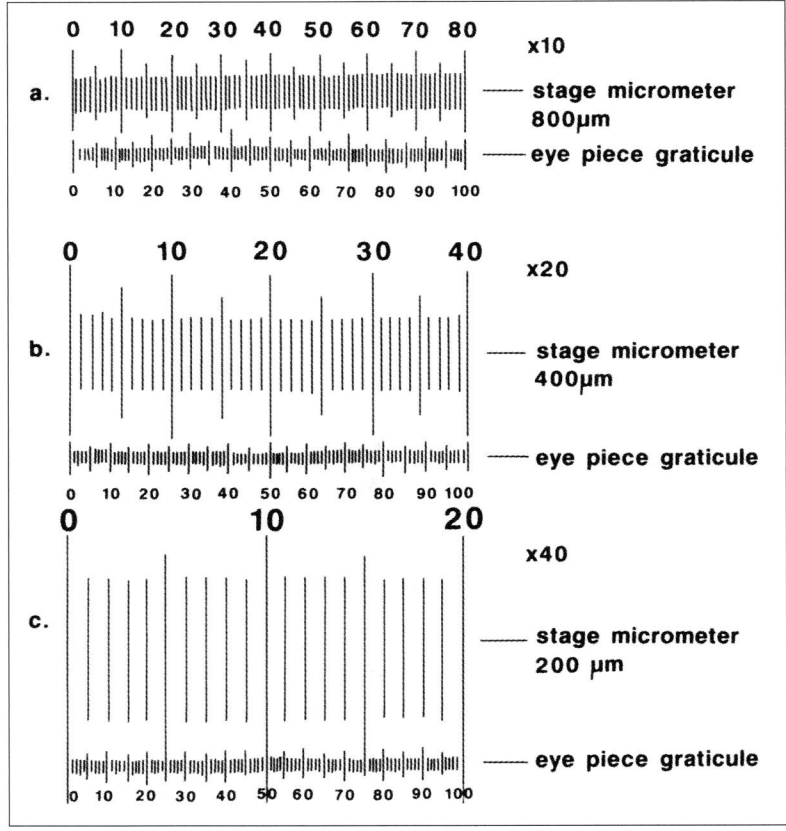

1–10 and not back to front. If the scale is back to front, the disc is upside down and the process needs to be repeated to insert the disc the right way up. A square graticule used for counting should be inserted in the same way as for calibration, similarly checking that any numbers and letters read correctly.

4.2 Calibrating an eyepiece graticule scale

1. Place the stage micrometer on the microscope stage like an ordinary slide with the cover slip uppermost, locate the micrometer stage scale using a low power objective and then use the objective for which the eyepiece is to be calibrated.

2. Move the micrometer scale so that zero marking coincides with the zero on the eyepiece graticule scale.

3. Make a note of the reading on the micrometer scale opposite the markings on the eyepiece graticule scale. Fig. 5.5c shows that 50 small eyepiece graticule subdivisions are the same length as 10 stage micrometer divisions. Since each stage micrometer division represents 10 µm, 50 small eyepiece divisions represent 100 µm (10 x 10 µm), therefore one small graticule scale gives measures 2.0 µm (100 µm ÷ 50).

4. The calibration for each magnification should be recorded on a card or label kept on or near the microscope for easy reference.

4.3 Measuring

To measure a pollen grain, spore or other structure

1. Move the eyepiece graticule so that the scale lies alongside or across the specimen at the same angle as the structure by turning the eyepiece in its socket.

2. Measure the object using the eyepiece scale graticule.

3. Use the calibration already obtained (from section 4.2) to calculate the actual measurement. When using the x40 lens if the length of the object measured 75 divisions on the eyepiece scale graticule the length would be 150 µm (75 x 2 µm).

As with all biological material, pollen and spores are not uniform in size but show a characteristic range of sizes for each species. The average, and mean size can be obtained by measuring many specimens.

Microscopic measurement is most accurate in the centre of the field and also only considered accurate to the nearest 0.5 µm.

4.4 Calibrating with an eyepiece square graticule

1. Place the stage micrometer on the microscope stage and locate the micrometer stage scale by using a low power objective and then use the objective for which the eyepiece is to be calibrated.

2. Move the micrometer stage so that the zero marking is over a square side line and parallel to side lines of the squares.

3. Make a note of the stage micrometer reading of a number of squares of the eyepiece graticule (the number of squares used will depend on the type of graticule). e.g. Fig. 5.6b shows that 5 squares measure 40 small micrometer divisions. Since each small micrometer division represent 10 µm, 40 divisions represent 400 µm (10 µm x 40), therefore 1 square measures 80 µm (400 µm ÷ 5).

4. The calibration for each magnification should be recorded on a card kept with the microscope or on the wall behind it.

Figure 5.6
Calibrating an eyepiece square graticule against a stage micrometer with different objectives (BAF, 1995, with permission).

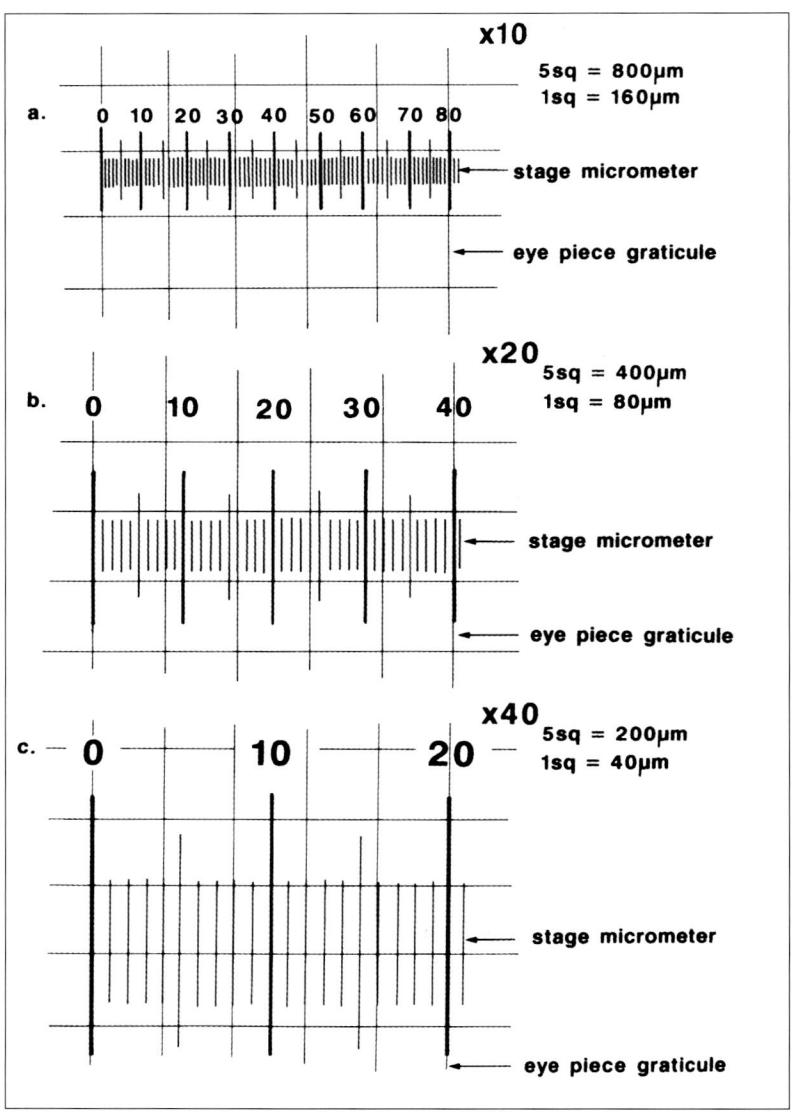

CHAPTER 6

Pollen and spore counts

1. Introduction

As already described in Chapter 3 airborne particles can be collected by many different methods, but this chapter deals only in detail with counting samples collected by the Burkard and whirling arm traps. The number of particles per cubic metre of air is the standard method of reporting the air spora.

All samples of airborne particles contain so much diversity that it is necessary to decide exactly what should be counted before the start. If a general ecological study is being made then a good look around the slide will give an idea of the categories that could be counted. Always have an open mind and be conscious of the local topography, ecology and weather conditions.

The deposit on the slide represents a measured volume of air as the rate and length of sampling time is known. It is too time-consuming and unnecessary to count every particle, so a known proportion of the area is counted and a correction factor used to calculate the concentration of particles in the air. Individual fields, or traverses across or along the length of the slide, can be counted (Figs. 6.2 and Fig. 6.6).

1.1 Examples of slides

The colour plates 1–3 are microphotographs of trap slides from various spore traps to give some idea of the diversity of the air spora found in different habitats. Most photographs were taken with a x20 lens but figures c–f on Plate 3 were taken with a x40 lens. On all slides there are many inorganic particles of various origin but they are not mentioned for every slide.

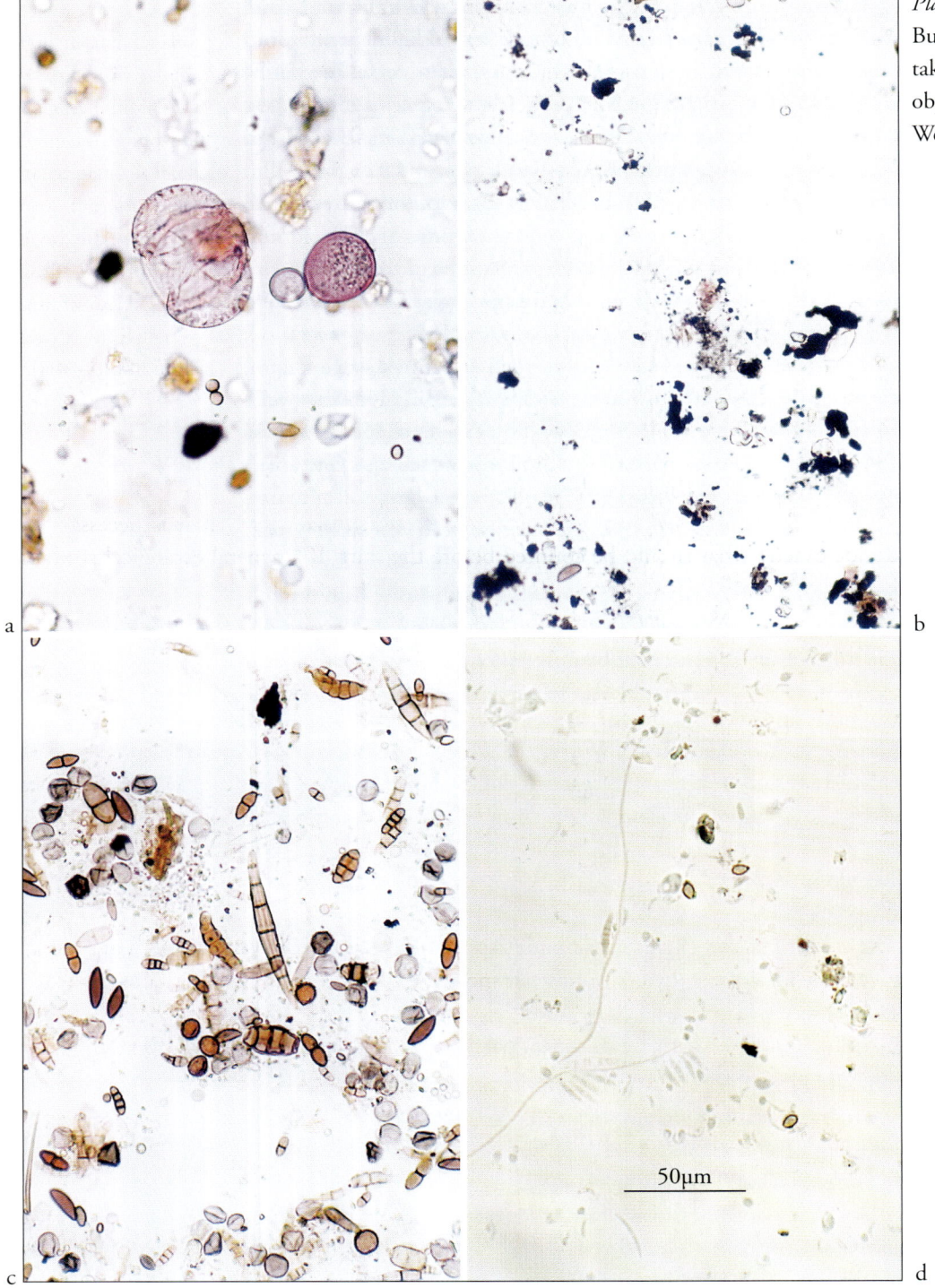

Plate 1
Burkard trap slides taken with a x20 objective (Jonathan West).

Plate 1, Burkard trap slides taken with a x20 objective

a) Burkard trap used for the pollen count at Worcester, UK, 14.06.2004, stained with safranin for easy recognition of pollen grains. Summer day showing grass and pine pollen, also *Cladosporium* and *Ustilago* spores.
b) Burkard trap used for the pollen count at Worcester, UK, 13.10.2004, stained with safranin. Damp autumn evening showing particle groups with some spores deposited from mist droplets.
c) Burkard trap sampling at 1.5 m above ground in a banana plantation, Costa Rica, 19.05.1992. Many single, two celled and multicelled fusiform ascospores are present, also conidia including a spore of *Pithomyces chartarum* (Burt *et al.*, 1999).
d) Burkard trap used in an oilseed rape crop at Rothamsted Research, UK, 24.05.1986. Summer early morning showing groups of 4 *Pyrenopezziza brassicae* ascospores, basidiospores, *Tilletiopsis* and *Sporobolomyces* spores (Lacey, M. *et al.*, 1987).

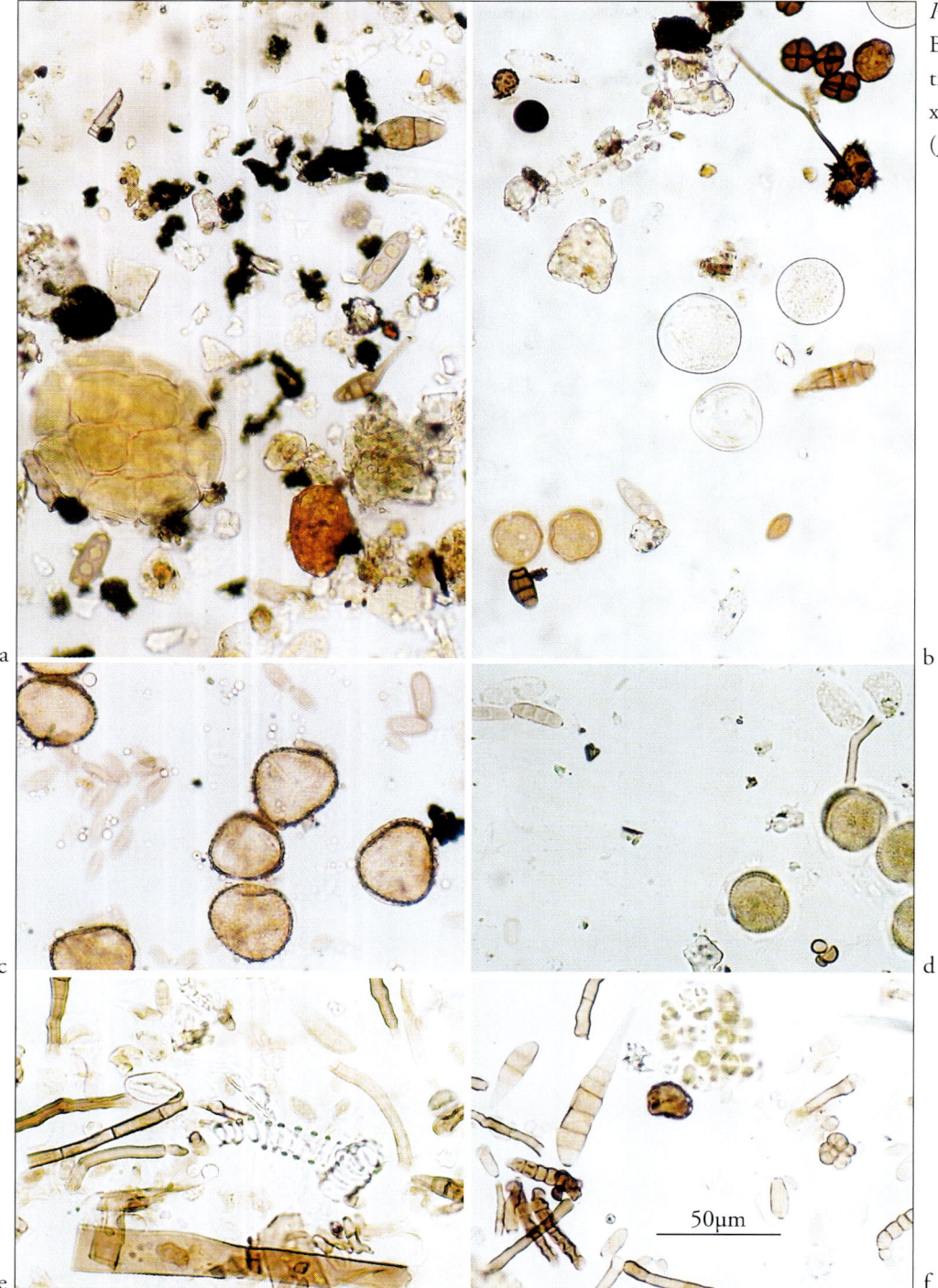

Plate 2
Burkard and travel trap slides taken with a x20 objective (Jonathan West).

Plate 2. Burkard and travel trap slides taken with a x20 objective

a) Travel trap running for 10 hours during a hot day near a road in Alice Springs, Australia, 21.12.2000. *Acacia* pollen and spores of *Alternaria* and *Drechslera* are present with large numbers of soil, rubber and other particles.

b) Travel trap running for 4 hours during a warm dry evening at Tallegalla, near Brisbane, Australia, 28.02.1997. Grass pollen and spores of *Spegazzinia lobulata*, *S. tessartha*, *Epicoccum*, *Alternaria*, *Drechslera*, *Nigrospora*, *Pithomyces chartarum* and rust are present.

c) Burkard trap near a stand of bracken at Rothamsted Research, UK, 01.10.1990. Spores of bracken (*Pteridium aquilinum*) and *Cladosporium* sp. are present (Lacey, M. and McCartney, 1994).

d) Burkard trap in an oilseed rape crop at Rothamsted Research, UK, 04.05.1987. Oilseed rape (*Brassica napus*) pollen and spores of *Cladosporium*, *Blumaria*, *Ustilago* and *Peronospora* are present.

e) Burkard trap near harvesting at Rothamsted Research, UK, 13.08.1985. Showing an Umbellifer pollen, pro-xylem, hyphal fragments, and spores of *Cladosporium* and *Alternaria*.

f) Burkard trap near harvesting at Rothamsted Research, UK, 13.08.1985. Showing a group of *Gloeocapsa* algae, hyphal fragments, *Alternaria*, *Cladosporium*, *Epicoccum*, and *Botrytis* spores.

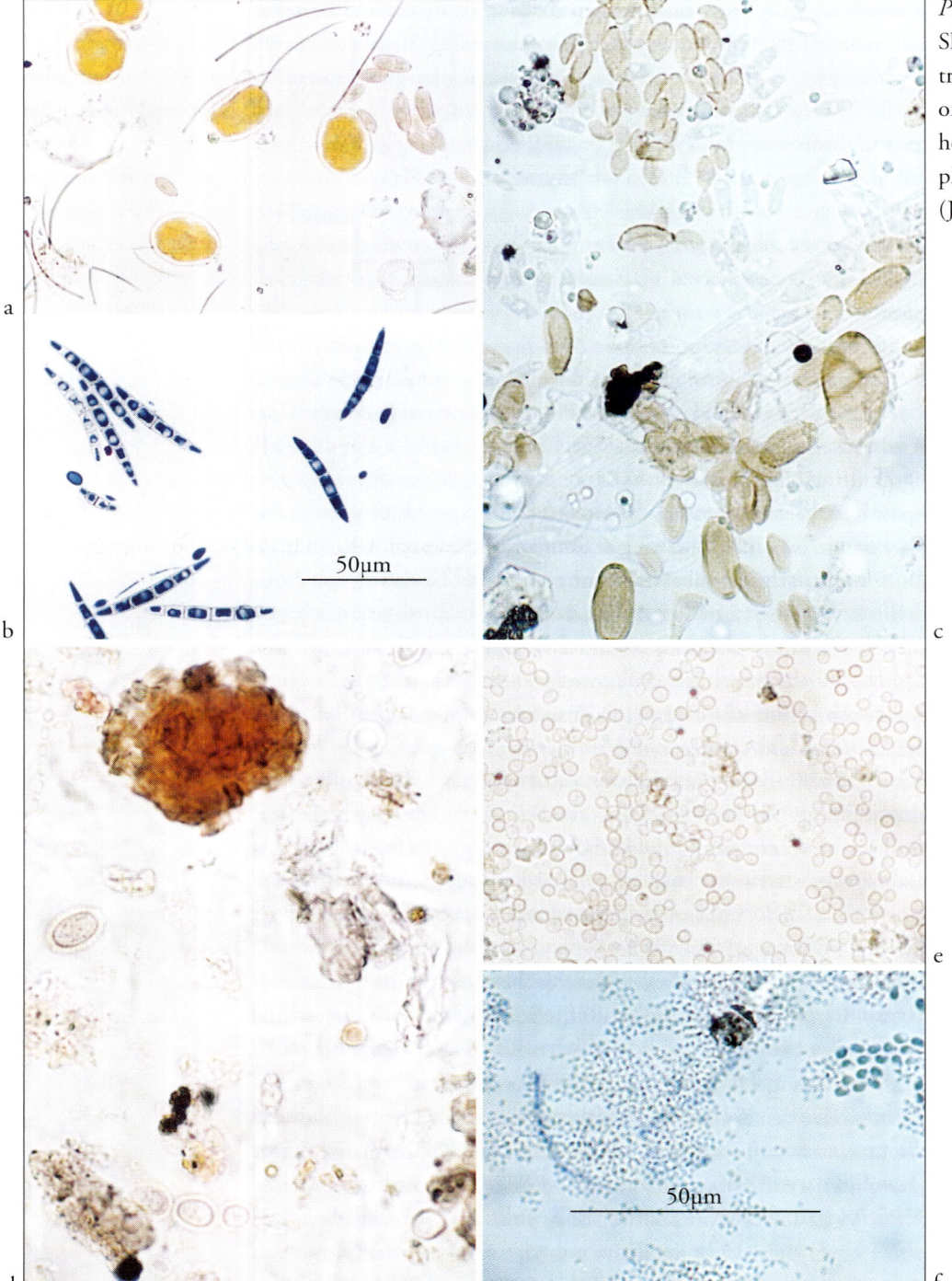

Plate 3
Slides from various traps taken with a x20 or x40 objectives, health hazards for plants and man (Jonathan West).

Plate 3. Slides from various traps taken with x20(a-b) or x40(c-f) objectives, health hazards for plants and man

a) Whirling arm trap in a field of wheat at Rothamsted Research, UK, 21.05.2003. Showing insect hairs, yellow rust (*Puccinia striiformis*) uredospores, *Cladosporium* conidia and a coiled filamentous spore. (West and McCartney, 2003).

b) Whirling arm trap in a field of oilseed rape at Rothamsted Research, UK, 23.10.1999. Showing ascospores of brassica canker (*Leptosphaeria maculans*) stained with trypan blue (West *et al*, 2002).

c) Whirling arm trap in a field of sunflowers at Rothamsted Research, UK, 05.07.1998. Showing part of an insect scale, conidia of *Cladosporium* and ascospores of stem rot (*Sclerotinia sclerotiorum*) stained with trypan blue (McCartney and Lacey, M., 1999).

d) Burkard trap in a students room, Worcester, UK, 01.12.2000. Showing a housedust mite faecal pellet, skin scales with bacteria, *Aspergillus* and smaller *Penicillium* type spores and other debris.

e) Cascade impactor, (stage 3) in a cork factory in Portugal, 17.03.1972. showing spores of *Penicillium* (mainly *P. frequentens*), a cause of suberosis in man (Avila and Lacey, J., 1974).

f) 7 stage impactor, (stage 4) at a mushroom farm, UK, 1988. Showing some *Penicillium* type and many actinomycete spores (Crook and Lacey, J., 1991).

1.2 Choosing the right magnification and trace width

When first looking at a new set of slides start with a low magnification and have a good look around to see the largest objects. Move up the magnifications until a suitable one is reached. If a general count is required then one enabling counts of the small particles must be used. When pollen and large spores are to be counted a magnification of x20 would be adequate, but if small fungal spores, bacteria or actinomycetes are being counted x40 is needed. The following Table gives an idea of possible uses of objectives in aerobiological studies, higher objectives may be needed on some microscopes.

Table 1. Use of objectives in aerobiology

Objective	Use
x10	general look
x20	counting pollen and large spores
x40	counting pollen and spores
x60	counting small spores
x100	looking at fine detail

An eyepiece square graticule is used to determine the width of the traverse (Fig. 6.1). The traverse width should always be several times wider than the largest particles being counted to prevent overestimating concentrations. Remembering that the edge of the field of vision is not as clear as the centre, 5 or 7 squares wide would be a good choice. If the trace contains very many small spores, e.g. hundreds of bacteria or actinomycete spores per square, then 3 or even only 1 square width should be counted.

The traverse width for all counts is calculated by multiplying the number of squares by the known calibrated width of each square.

Whilst counting, the focus has to be constantly adjusted to find all the particles and to aid identification. Stains can be used to emphasise certain particles, e.g. safranin for pollen and cotton blue for hyaline fungal spores. A bank of mechanical counters enables easier recording of the most common particles.

For any series of counts it is necessary to always follow the same procedures so that direct comparison can be made between sets taken at different times.

1.3 Counting conventions

Once the width of the traverse is decided it is very important to be consistent in which particles are to be counted as some will lie across the edge of the counting area. A pollen or spore could be counted if more than half of it was within the counting area and

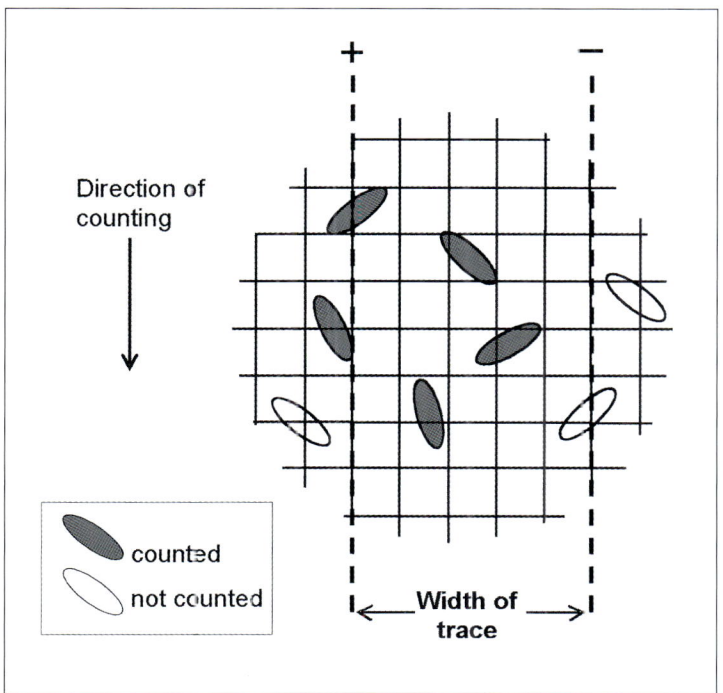

Figure 6.1
Counting conventions for a transverse traverse of a Burkard trap slide: particles completely within the width of the trace or touching the '+' edge are counted and those outside or touching the '−' edge are not counted.

excluded if more than half was outside, but it is easier to be consistent to include only those touching the + edge and ignoring those touching the − edge (Fig. 6.1). The width of the 5 square traverse using the same microscope as in Chapter 5 would be 800 μm at x10, 400 μm at x20 and 200 μm at x40 (Fig 5.6).

The microscope slide is moved up or down in steps for each transverse traverse and care must be taken not to count any particle twice after moving the slide.

Figure 6.1 shows a transverse traverse but if a longitudinal one is counted then the squares counted would be grouped horizontally, i.e. the slide moved from left to right and the particles counted would be either those on the edge at the top or bottom of the traverse.

2. Counting pollen and spores from a Burkard trap

2.1 Introduction

Longitudinal traverses from a Burkard trap can be used to give average daily concentrations or transverse traverses to give hourly concentrations. The slide in Figure 6.2 shows a 24 hr. deposit from a 7-day Burkard trap situated by a stand of bracken in August 1990 (Lacey, M., and McCartney, 1994). The trap was changed at 09.00 hr GMT to enable direct comparison with meteorological data (Hirst, 1952). It is necessary to

Figure 6.2
Burkard trap slide showing the position of a longitudinal and transverse traverses.

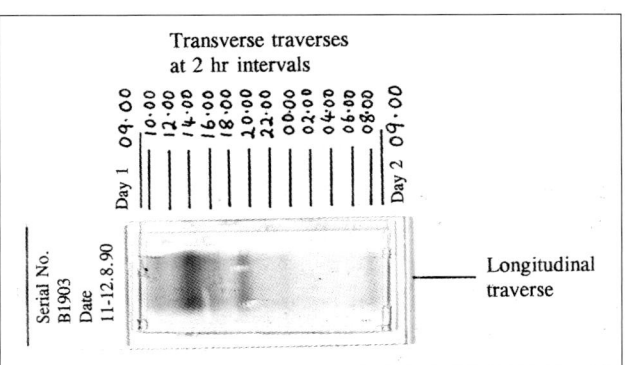

change the slide of a 24-hr. trap (Fig. 4.1c) earlier for pollen counts when counting deadlines have to be met.

2.1.1 *Deposition of particles*

Before deciding how to count particles caught from a Burkard trap it helps to understand how they are deposited. The slit of the orifice through which the particles enter the trap is 2 mm wide (Fig. 6. 3a) and it passes over the tape or slide at 2 mm an hour, therefore it takes an hour to pass from one edge of the deposit area to the other (Fig.6. 3b). Although the movement of the tape or slide is continuous, each six minutes is represented by a horizontal line on the diagram, this allows a visualisation of the way the deposit is built up. Each deposit of one hour overlaps with the next.

Figure 6.3
Deposition of particles on the trapping surface over a one-hour time-period, each horizontal line is equivalent to 2 mm length of trace or 1 hour exposure (BAF, 1995, with permission).

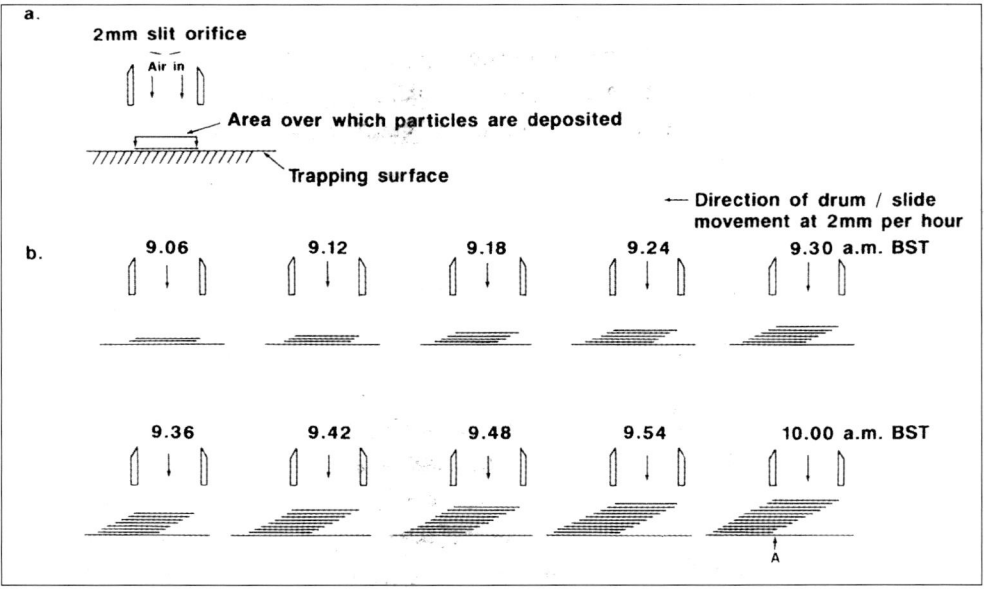

Figure 6.4
a) 24-hour trace from a Burkard trap showing 12 transverse traverses. b) deposition of small and large particles across a spore trap slide at different wind speeds (BAF 1995, with permission).

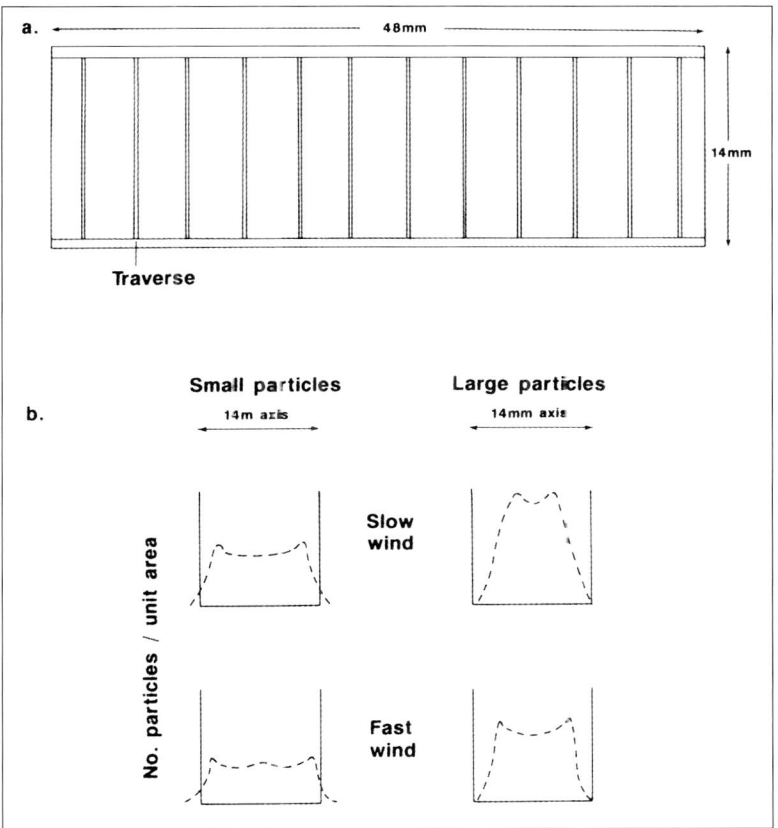

As shown in Figure 6.4b the particles are not deposited evenly across the 14 mm width of the trace (Hirst, 1952). There are twin zones on each trace where the deposition is heavier; this differs for small and large particles, and also at different wind speeds. This effect can be seen clearly in Figure 6.2 where the deposit is particularly dense around 14.00 hr.

2.1.2 Daily counts – longitudinal traverses

Longitudinal traverses give the average concentration of particles in the air over a period of 24 hours. A series of daily counts can show the changes in the air spora over a period of a week, month, season or year.

As seen in Figure 6.4b the particles are not evenly deposited across the trace and consequently only one longitudinal traverse will not give an accurate figure for the air spora concentration.

Daily counts give a good idea of the contents of the air. In plant pathology, daily counts are very important to show the presence or absence of spores in the air. When they are present, detailed hourly counts can be made to study the release and dispersal of the pathogens in relation to the meteorological conditions. Similar methods can be used for studying dust present in both indoor and outdoor work environments.

2.1.3 Hourly counts – transverse traverses

Counting twelve transverse traverses (Fig. 6.4a) rather than a number of longitudinal ones allows all relevant particles along the 14 mm axis to be counted and takes account of the unevenness (Fig. 4b). It also allows diurnal concentration patterns to be determined and compared with weather conditions.

Although unlikely with pollen grains, a sudden burst of fungal spores, perhaps after a summer thunderstorm, can sometimes be missed between two-hourly transverse counts although they would be detected using a longitudinal scan.

2.2 The daily pollen and spore count

To meet press and news deadlines in the UK, it is usual to change the traps at 9 a.m. British Summer Time (BST: 0800 Greenwich Mean Time, GMT) and then to count the grass pollen grains that occur in twelve evenly spaced transverse traverses of the slide. The first of these traverses is at 10 a.m. BST (0900 GMT) on the day before the count is made and the twelfth at 8 a.m. BST (0700 GMT) on the day of counting. The pollen and/or spores in each of the twelve traverses are added and then multiplied by a Correction Factor to give the 'Daily Pollen Count'.

Full records of the microscope set-up, dates, times and counts should be entered on a standard counting sheet (e.g. Appendix 4).

2.2.1 The positioning of transverse traverses

Through the first hour of trapping, the trace represents less than one full hour's trapping. Near the start of deposit, there will have been only a few minutes trapping, e.g. at 9.06 a.m., and a traverse near the start of the trace would only represent the number of particles trapped in 6 minutes, a notional tenth of the catch for an hour. Point A, after one hour, is the first point at which there is a 'complete' hour's deposit and the earliest a representative traverse could be made.

Pollen and spores caught in any one-hour period may spread over 4 mm of trace and the catch at any one point represents the mean concentration of airborne particles over one hour. It was at the exact centre of the orifice at 9.30 a.m. Thus 9.30 a.m. best represents the time between 9 a.m. and 10 a.m. since it is the point to which all the 'overlapping traces' for one hour contribute. Transverse counts are usually made on the hour so that 10 am is the first complete 'on-the-hour' traverse, representing the mean concentration between 9.30 a.m. and 10.30 a.m. At the end of a 24 hr. trace, the 8 a.m. traverse is the last complete 'on-the-hour' count (Fig. 6.5a-c)

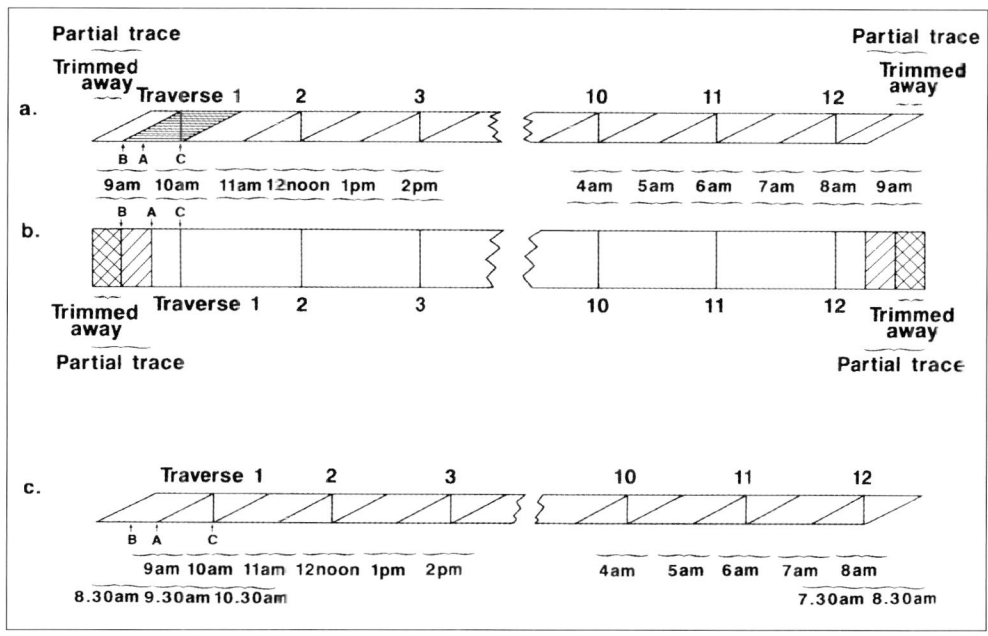

Figure 6.5 Positions of traverses in relation to times of deposition of pollen or spores (BAF, 1995, with permission).

The 10 am traverse is 3 mm from the beginning of a 24-hour trace. If the trace is on a slide, 3 mm should be measured from the beginning of the trace, using the vernier scales on the microscope stage. If the trace was formed on the Melinex tape, the first millimetre of the first and last segments should have been trimmed away on mounting to allow the tape to fit beneath a 50 mm cover slip and 2 mm should be measured from the trimmed edge of every segment of tape to the position of the first traverse.

2.2.2 Counting and recording the daily counts

1. Using a low power objective, find the edge of the trace, and note its position using the vernier scales on the microscope's mechanical stage (Fig. 5.2).

2. With traces deposited directly onto slides, measure 3 mm in from the beginning of the trace to the position of the first traverse. Although the drum rotates at 2 mm per hour, an additional one mm is discarded at the start of the trace (due to spore deposition), hence 3 mm into the trace corresponds to one hour after starting. If the trap is changed at 9 a.m. BST, this will represent 10 a.m. BST: if changed at other times, the position or the recorded time will have to be adjusted accordingly.

3. Traces on Melinex tape should have the first millimetre trimmed away from the first and last segment so that the position of first traverse of all segments is 2 mm from the beginning of the tape.

4. Place the x40 objective (or the one being used) in position and adjust the microscope.

5. If using the counting sheet (Appendix 4), start counting from the bottom of the trace along the first traverse. Otherwise, keep a note of the direction the slide is moving under the microscope as it is all too easy to look away and then forget which direction is being followed. An arrow pointing up or down is quick and easy to understand.

6. Observe the counting conventions (Fig. 6.1) to determine whether a grain is within the traverse or not.

7. Note the traverse width in number of graticule squares. This will depend on the size and number of spores or pollen. For pollen, ten graticule squares at x40 magnification have been recommended (BAF, 1995) but fewer can be counted with fungal spores. The best traverse width will also vary with each microscope set up.

8. Record the number of grass pollen grains counted along the traverse.

9. Move the slide 4 mm, i.e. 2 hours, using the vernier scales.

10. Count for the next traverse and record the total.

11. Add the twelve counts to give a total for the whole slide. Multiply this total by the Correction Factor appropriate to the traverse width that is being used.

2.2.3 Pollen and spore concentrations and correction factors

The pollen grains or fungal spores counted on a spore trap slide form only a small sample of the total number which have been trapped. Therefore the number counted must be 'corrected' to give the actual concentration of grains per cubic metre of air.

a) Total the number of grass pollen grains or spores in each of the twelve traverses to get the daily total count (N).

b) Multiply this total by the Correction Factor (CF) for your microscope/lens combination using:

$$N \times CF = N \times \frac{0.28}{\text{width of one traverse (mm)}}$$

The result of this calculation is the daily mean concentration of pollen grains or fungal spores per cubic metre.

2.2.4 Calculating a correction factor

To calculate a correction factor to convert the total counts of different pollen or spores on the twelve traverses to concentrations per cubic metre (m^3), the exact size of the traverses counted must first be measured (in mm) and the area (in square mm or mm^2) calculated (the width of a traverse in mm can be derived by dividing the width in µm by one thousand, i.e. 250 µm = 0.25 mm). The proportion that this forms of the total area of the spore trap is determined to give a multiplication factor which converts the number of spores on the area counted to that on the whole trace. This is then divided by the volume of air sampled in 24 hr to give the final correction factor.

a) The total area of a 24 hr spore trap trace is:

   48 *mm* x 14 *mm* (Fig.6.4a)

b) The area of one traverse is:

   (14 *mm* x *width of traverse in mm*) mm^2

c) The total area of the 12 traverses counted is:

   (14 *mm* x *width of one traverse in mm* x 12) mm^2

d) The fraction of the total area counted, i.e., occupied by the twelve traverses is therefore:

$$\frac{14 \text{ mm x width of one traverse (mm) x12}}{48 \text{ mm x 14mm}}$$

e) If N grains or spores are counted in twelve traces, the total number deposited on the slide is:

$$N \times \frac{48 \times 14}{14 \times \text{width of one traverse} \times 12} \text{ grains}$$

f) Concentrations of pollen grains are usually expressed in numbers per cubic metre of air.
   1. In 24 hours, a standard spore trap samples 14.4 cubic metres of air (10 l min^{-1} x 60 min x 24 hr.), depositing its content of pollen and spores on the trapping surface.
   2. The calculated total of pollen grains or fungal spores on the slide must therefore be divided by 14.4 to give the average concentration of pollen grains or fungal spores per cubic metre of air over 24 hours.

g) The total calculation to convert the raw count to a concentration, in pollen grains or fungal spores m^{-3}, is thus:

$$N \times \frac{48 \times 14}{14 \times \text{width of one traverse} \times 12 \times 14.4}$$

h) This can be reduced to:

$$N \times \frac{0.28}{\text{Width of one traverse (mm)}}$$

i) The correction factor is the expression:

$$\frac{0.28}{\text{Width of one traverse (mm)}}$$

j) Having calculated the correction factor (CF), counts are converted to concentrations m^{-3} using:

$$N \times CF$$

The 'Correction Factor' is specific to the microscope and lens combination for which it is calculated. It can be calculated at the start of the pollen or spore counting season and then used throughout, providing conditions such as microscopes or lenses do not change.

Three examples for different traverse widths:

a) For a traverse 250 μm or 0.25 mm wide, the C.F. is: **0.28 ÷ 0.25 = 1.12**

b) For a traverse 125 μm or 0.125 mm wide, the C.F. is: **0.28 ÷ 0.125 = 2.24**

c) For a traverse 400 μm or 0.4 mm wide, the C.F. is: **0.28 ÷ 0.4 = 0.7**

d) In Fig. 5.6c a field from a microscope in which 5 eyepiece graticule squares measure 200 μm is illustrated.

The correction factor for that microscope with a ×40 objective, if a traverse 10 squares wide is counted, would be:

0.28 ÷ 0.4 = 0.7

If 150 pollen grains were counted, this correction factor would give a mean daily concentration of 105 grains per cubic meter. The correction factor for a traverse 0.25 mm wide, as shown in the first example above is 1.12 which with the same number of spores would give a mean daily concentration of 168 spores m^{-3}.

It is thus extremely important to measure the width of the traverse accurately to obtain an accurate correction factor for the microscope eyepiece and objective being used.

3. Counting pollen and spores from a whirling arm trap

3.1 Counting spores on tape sections

The width of rotating arms and therefore the section of tape peeled off and mounted on a slide varies depending on materials used in their manufacture but typically a 1.6 mm-wide tape section fits into the field of view of many microscopes at a magnification of around x10. This means that for large, obvious spores e.g. *Alternaria*, it is possible to scroll down the length of a tape section to count spores using a microscope. However, it is more usual, especially when counting small spores or when the spores are very numerous, to count a certain number of transverse traverses of each tape section (Fig. 6.6 shows 5 traverses per tape section). If the number of specific particles is very high, particles in only small sub-sections of each strip may be counted, but the same positions should be counted for each trap, where comparisons of multiple trapping positions are to be made, because the distribution of spores can vary with location on the trap tape.

Figure 6.6
Whirling arm trap slide showing position of counting traverses

3.2 Calculating particle concentration

The particle concentration per m^3 of air, collected during a sampling period, can be calculated by dividing the number of particles counted per arm (mean of 4 tape sections) by the volume of air sampled during the sample period. The volume of air (V) sampled by each arm of the trap is calculated using the equations for air sampled per minute multiplied by the duration of trapping in minutes i.e.

$$V = \pi.D.W.L.S.T.\ 10^{-3}\ \text{litres}$$

where D is the outer diameter (in cm), W the width (of collecting surface in cm), L is the length (of the collecting surface in cm), S is the rotation speed in rpm and T is the period of trapping in minutes.

For example if: D = 7.8 cm, W = 0.16 cm, L = 5.9 cm, and S = 3600 rpm; then each arm samples: 83.27 l min^{-1}.

If the trap were run for two hours then the volume of air sampled per minute is multiplied by 2 x 60, so the total volume sampled per arm in this example is

83.27 x 2 x 60 = 9992 litres, or 9.992 m^3.

If the number of particles in only a portion of the arm area were counted, the number is scaled up to give the number of particles per whole arm, before calculating the number of particles per m^3 of air sampled. For example if five traverses were made on each of the four quarters of the arm-tape and each traverse had a field diameter of 280 µm, the area observed is 20 (the number of traverses per arm) x 1.6 mm (the width of the arm) x 0.28 mm (field diameter), which equals 8.96 mm^2. Since the area of the whole arm is 1.6 x 59 mm, i.e. 94.4 mm^2, the number of particles per whole arm is calculated from the number counted in 8.96 mm^2 x 94.4 ÷ 8.96.

CHAPTER 7

Identification

1. Introduction

Visual identification by light microscopy of airborne particles caught on a sticky surface is not easy (Allitt, 1979), but it still remains a good way of examining the whole air spora. Many pollen and spore atlases have good drawings or photographs but often many different magnifications are used. Gregory (1973) maintained that it was important to have all illustrations at the same magnification to enable easy comparison. The paintings in this book are all x1000 and give a selection of particles that can be found in the air. Identification to the level of species is often not possible. Many people looking at slides of particles deposited from air are often very specialized in a certain subject and therefore are not able to identify other kinds of particles present. Glancing through these plates will often give a hint of identification and it is then easier to look at different books or ask workers with expertise in that area. Small particles can be found as contaminants on many microscopic slide preparations for example a pollen grain or a fungal spore can be found on a histology preparation. The following plates could help such contaminants to be identified.

When trying to identify an object it is usually necessary to change focus constantly to look closely at its size, shape, density, colour, thickness and texture of its wall or edge, and its contents. Higher magnifications should be used to look at smaller particles but always be conscious of the size by checking with the calibration on the eyepiece graticule in the microscope. It is useful to build up a good collection of specimens for reference.

2. Particle types

Microscopic particles in the air can be of any constitution but this book is mainly concerned with biological particles. Small particles that are able to enter the lungs can cause problems to health for example asbestosis is caused by the inhalation of asbestos fibre. Examples of the diversity of particles in the air at any period of time can be seen in Plates 1-3 (chapter 6).

2.1. Pollen

Pollen grains tend to be spherical or nearly so, pale in colour and are the male gametophyte of seed plants. They are not very dense and contain a granular protoplasmic mass. The pollen wall, the exine, is multi-layered and often ornate with pores and fissures. The exine contains sporopollenin, which is resistant to decay, and hence pollen may last for thousands of years in the right conditions e.g. in peat. Pollen dispersed by wind tends to be smaller than those dispersed by insects, but sometimes have large air sacs (Pl. 7.1-3). Grass pollen is the most common type of pollen that is counted regularly, and typically has a single pore, which can be orientated in any aspect. The painting of *Phleum pratense* shows the pore in cross section whilst that of *Alopecurus myosuroides* shows it on the upper surface (Pl. 4.1 and 6).

2.2. Fungal Spores

Spores are classified by their mode of production. Fungal spores are very varied, single- to multiple-celled, hyaline with thin walls to dark with thick walls. Basidiospores are produced from a spore-bearing structure called a basidium, produced by all basidiomycetes such as rusts and smut fungi and the macro-fungi that produce mushrooms, toadstools and brackets. They have a scar of attachment and are mostly longer than broad and pale to dark, the wall is smooth or sometimes patterned (Pl. 9.1-50). Ascospores are produced in eights in an ascus and are released explosively by changes in water pressure by drying after rain. They have no scar of attachment, are elliptical to elongated, single or multi-celled, hyaline to dark (Pl. 8.1-46) and are often found in the air in groups of up to eight (Pl. 8.19 and 20). Many spores are produced asexually and are called conidia, they are single- to multi-celled, often coloured and may be quite large, the scar of attachment can usually be seen, and they are often released dry by the wind (Pl. 10.1-51, Pl. 11.1-27). Small spherical spores such as those of the genera *Aspergillus* and *Penicillium* are very difficult to identify to species level. Spores that are splashed tend to be hyaline with thin walls, e.g. *Rhynchosporium secalis* (Pl. 10.34).

2.3. Other plant material

Bacteria and actinomycete spores are very small and hyaline, more or less spherical and down to 1μm in diameter (Pl. 12.2-6). Lichens are composed of fungi and algae in symbiotic association as seen in the painting of the lichen soredium *Cladonia* (Pl. 11.38), the fungal constituent is often an Ascomycete e.g. *Xanthoria parietina* (Pl. 8.21). Algae grow in damp areas but can get airborne e.g. *Gloeocapsa* (Pl. 11.37). Diatoms are microscopic algae with delicately sculptured siliceous cell walls which are very refractive (Pl. 11.34-36).

Ferns are distributed worldwide and usually grow in damp areas, e.g. valleys and tropical rainforests, spores are produced in sori which rupture to release them (Pl. 7.6-14). Mosses also grow in any damp places, their spores are produced in capsules and can be found in the air (Pl. 11.28-33). Plants become broken and pieces are blown around in the air, plant hairs become detached and even structures such as pro-xylum (Pl. 12.10) can be found.

2.4. Animal material

Small insects, such as thrips, are often found in suction traps. They are removed before mounting the slide but often scales, hairs and other parts are left behind (Pl.12.17-21). House dust mite faecal pellets and parts are known to cause allergenic reactions in many people. (Pl. 12.15 and 16). Cysts of small animals can also be found in the air (Pl. 12.14 and 22).

2.5. Inorganic material

Outdoor air samples contain many non-biological(abiotic) particles from soil and products of combustion (Pl. 12.23-32). The paintings are only of small examples to preserve space on the plate.

3. Useful references

Early studies of airborne pollen grains as a background study of inhalant allergen enabled Hyde and Adams (1958) to publish their book *An Atlas of Airborne Pollen Grains*. This was not long after the Hirst Spore trap had been developed and is extremely helpful in providing photographs, good descriptions and a key for identifying fresh pollen grains of 92 plant species found in Great Britain. The book *Atlas of Airborne Pollen Grains and Spores in Northern Europe* (Nilsson *et al.*, 1977) contains light microscope photographs and scanning electron micrographs of 69 pollen types and of 5 fern spores, with short descriptions and distribution maps in Scandinavia. *Pollen Analysis* (Moore *et al.*, 1991) deals more with pollen that has already been deposited from the air in lakes and peat. However it has good diagrams of pollen shapes and aperture types, a key for identification, and illustrations of 450 pollen types. The light microscope photographs are mainly at a magnification of x1000 and scanning electron micrographs range from x1000 to x8000. *Airborne Pollen and Fungus Spores* (Tilak, 1989) is useful for a reference of tropical pollen and spores.

Atlas of Airborne Fungal spores (Nilsson, 1983) has a very good introductory section showing the terminology used and the lifecycles of different groups of fungi, and illustrations of 87 species. The species descriptions produced by CABI Biosciences (previously CMI or IMI) are very useful for identification at species level. The books, *Genera of Hyphomycetes* (Carmichael *et al.,* 1980) and *Illustrated Genera of Imperfect Fungi* (Barnet and Hunter, 1998), have many good line drawings.

Many detailed line drawings of fern spores in the book A *Manual of the Spores of New Zealand Pteridophyta* (Harris, 1955) give a good idea of the diversity of the group. *An Atlas of Recent European Bryophyte Spores* (Boros *et al*, 1993) gives light microscope photographs mainly at x1000 and some scanning electron photographs. Illustrations of diatoms can be found in *An Illustrated Catalogue of Airborne Microbiota from the Maritime Antarctic* (Chalmers *et al.*, 1996) There is increasing interest in health and safety in the work place; *Microorganisms in Home and Indoor Work Environments* (Flannigan *et al.*, 2001) gives illustrations of 87 fungi and 3 actinomycetes that might cause health problems through allergy, toxicosis or infection (Samson *et al.*, 2001). *Identification of Pathogenic Fungi* (Campbell *et al.*, 1996) gives good drawings of pathogenic fungi; the spores of some become airborne. Many papers giving illustrations of pollen, fungal spores and other airborne particles have been published in journals such as *Aerobiologia* and *Grana*.

4. Paintings - x1000

For illustrating and identifying particles, colour paintings are preferable to photographs, as more than one plane of focus can be shown on each specimen. It is also much easier to have a large number of paintings on one plate. The following nine plates contain paintings by one of the authors (Maureen Lacey) and give a personal impression of the particles as seen down a light microscope. Paintings used from *The Microbiology of the Atmosphere* (Gregory, 1973) were made by using a camera lucida on an old upright monocular microscope with an oil immersion x100 objective. All other paintings were made using a modern binocular microscope with a drawing tube attachment, this is easier to use and must be set up accurately, using a stage micrometer, to give the required magnification of x1500 (Lacey, M., 1997). The outline of the particle is painted and then the drawing tube closed whilst the rest is painted, the tube is switched back as required to add further details. The original paintings were traced onto plates with a mounted needle to give a clean edge. For publication the plates were reduced and printed at x1000 so that a ruler can easily be used to measure the painting, 1 cm on the plate is 10 µm on the specimen.

Pollen was painted from fresh specimens (s) collected by workers in many countries. The Monocotyledons and Dicotyledons are arranged in order as in Engler's classification (Willis, 1960). Fungal spores and other particles were painted from specimens (s), cultures (c) or impaction traps (t). The spores are arranged in groups using the 8th edition of the *Dictionary of Fungi* (Hawksworth *et al.*, 1995), but the large group of Mitosporic fungi is arranged as in the 5th edition of the *Fungi Imperfecti* (Ainsworth, 1961). Common names have been given where possible. When there is a perfect and imperfect state painted they are cross-referenced (-). Square brackets [-] indicate a possible disease in plants or animals. The paintings on the Miscellaneous plate (Pl. 12) are very varied but often only show very small particles of a type, enabling as many as possible to be shown.

Many planes of focus were used for each painting. The majority of paintings show the cross section but some, that appeared more solid, do not. The paintings of pollen of *Casuarina equisetifolia* and *C. cunninghamiana* (Pl. 4.15 and 18) illustrate this point. Generally only one painting of each type is given, this is the best representative on the mounted slide available.

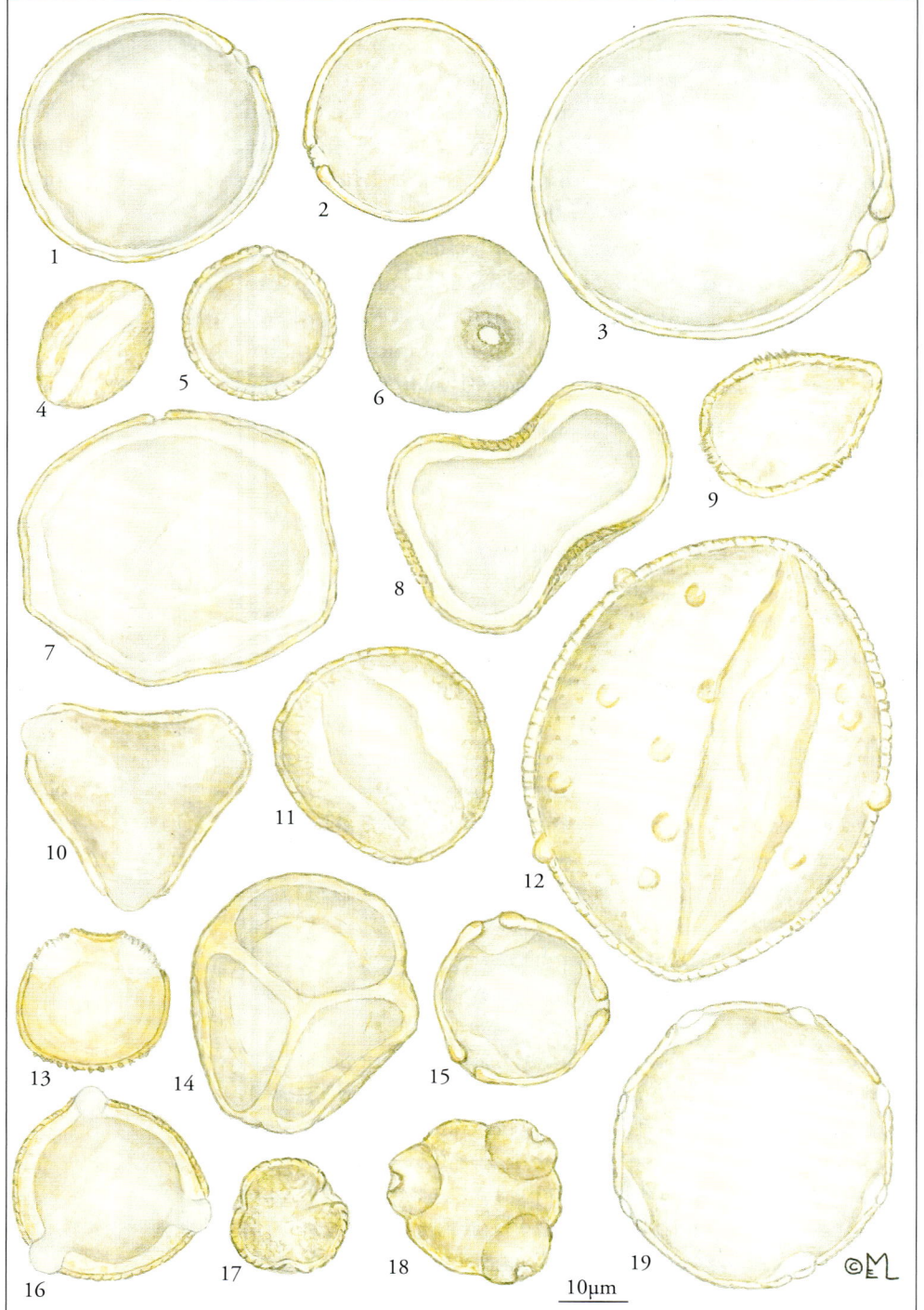

Plate 4
Pollen, Monocotyledons and Dicotyledons

Plate 4. Pollen - Monocotyledons and Dicotyledons

1 *Phleum pratense*, Timothy Grass, Gramineae, UK, s
2 *Paspalum dilatum*, Gramineae, Australia, s
3 *Triticum sativum*, Wheat, Gramineae, UK, s
4 *Phoenix sylvestris*, Date Palm, Palmae, India, s
5 *Sparganium erectum*, Bur reed, Sparganiaceae, UK, s
6 *Alopecurus myosuroides*, Black Grass, Gramineae, UK,
7 *Cocos nucifera*, Coconut Palm, Palmae, India, s
8 *Carex nigra*, sedge, Cyperaceae, Sweden, s
9 *Kyllingia* sp., Cyperaceae, Malaysia, s
10 *Elaeis guineensis,* Oil Palm, Palmae, Singapore. s
11 *Areca catechu*, Betel Nut, Palmae, India, s
12 *Borassus flavellifer*, Palmae, India, s
13 *Trachycarpus fortunei*, palm, Palmae, USA, s
14 *Luzula campestris*, Field Woodrush, Juncaceae, UK, s
15 *Casuarina equisetifolia*, Casuarinaceae, Singapore, s
16 *Garrya elliptica*, Garrya, Garryaceae, UK, s
17 *Salix caprea*, Goat Willow, Salicaceae, UK, s
18 *Casuarina cunninghamiana*, Casuarinaceae, Spain, s
19 *Juglans regia*, Walnut, Juglandaceae, UK, s

Plate 5
Pollen, Dicotyledons

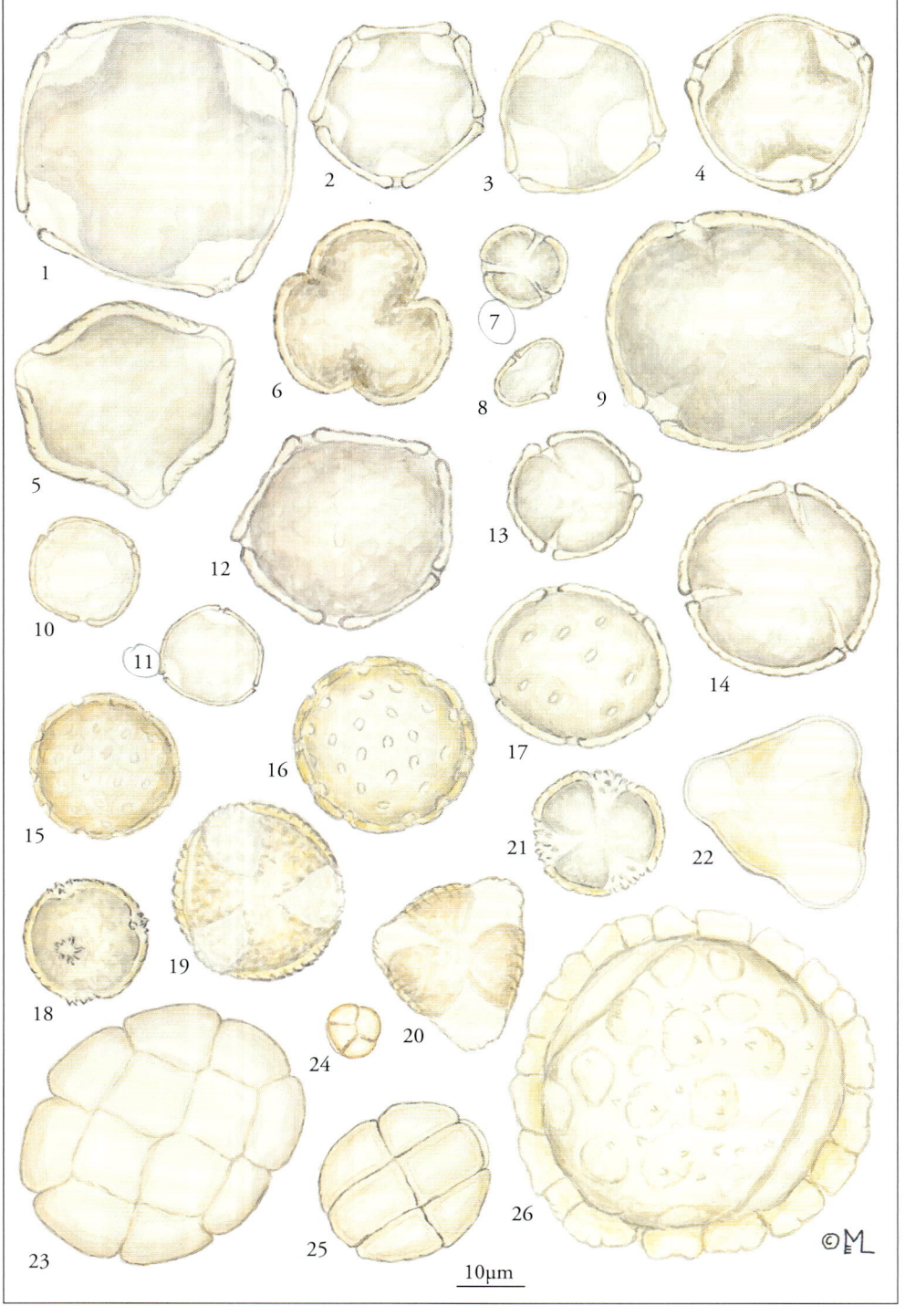

Plate 5. Pollen - Dicotyledons

1   *Carpinus betulus*, Hornbeam, Corylaceae, Sweden, s
2   *Alnus glutinosa*, Alder, Betulaceae, Sweden, s
3   *Corylus avellana*, Hazel, Corylaceae, UK, s
4   *Betula verrucosa*, Silver Birch, Betulaceae, UK, s
5   *Quercus robur*, Common Oak, Fagaceae, UK, s
6   *Quercus suber*, Cork Oak, Fagaceae, Spain, s
7   *Castanea sativa*, Sweet Chestnut, Fagaceae, UK, s
8   *Cecropia* sp., Moraceae, Venezuela, s
9   *Fagus sylvatica*, Beech, Fagaceae, UK, s
10 *Parietaria diffusa*, Urticaceae, Spain, s
11 *Urtica dioica*, Nettle, Urticaceae, UK, s
12 *Ulmus glabra*, Elm, Ulmaceae, UK, s
13 *Rumex acetosa*, Sorrel, Polygonaceae, UK, s
14 *Rumex crispus*, Curled Dock, Polygonaceae, UK, s
15 *Atriplex canescens*, Salt Bush, Chenopodiaceae, USA, s
16 *Chenopodiun album*, Fat-hen, Chenopodiaceae, UK, s
17 *Amaranthus viridis*, Amaranthaceae, India, s
18 *Thalictrum* sp., Meadow Rue, Ranunculaceae, UK, s
19 *Brassica napus*, Oilseed Rape, Cruciferae, UK, s
20 *Sorbus aucuparia*, Mountain Ash, Rosaceae, UK, s
21 *Platanus* sp., Plane, Platanaceae, UK, s
22 *Kalanchoe* sp., Crassulaceae, Venezuela, s
23 *Acacia auriculiformis*, Leguminosae, Singapore, s
24 *Mimosa pudica*, Leguminosae, Venezuela, s
25 *Mimosa arenosa*, Leguminosae, Venezuela, s
26 *Delonix regia*, Gulmohor, Leguminosese, India, s

Plate 6
Pollen, Dicotyledons

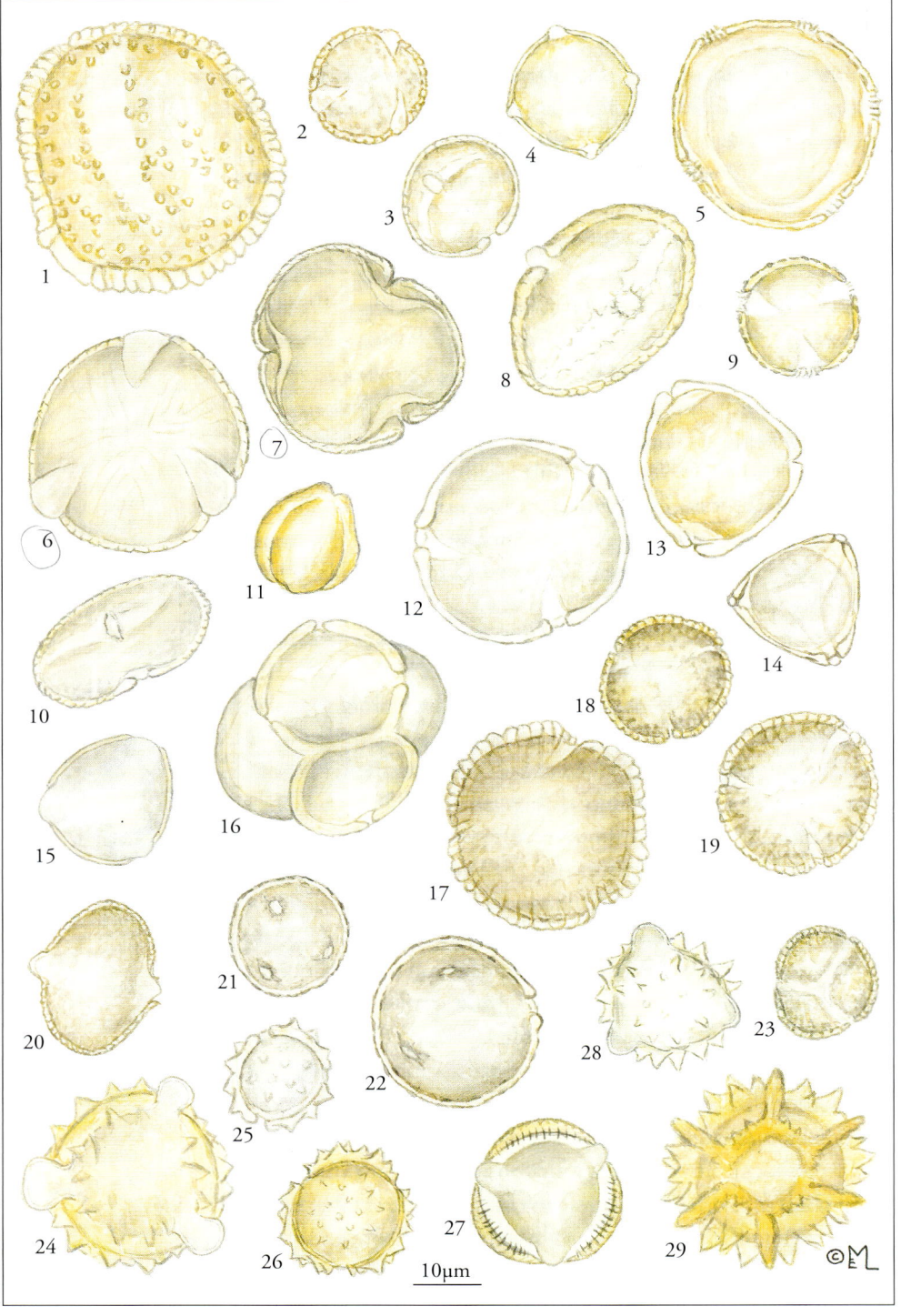

Plate 6. Pollen - Didotyledons

1 *Azadirachta indica*, Neam, Meliaceae, India, s
2 *Euphorbia hirta*, Euphorbiaceae, India, s
3 *Phyllanthus virgatus*, Euphorbiaceae, India, s
4 *Acalypha* sp., Euphorbiaceae, Venezuela, s
5 *Pistacia lentiscus*, Mastic Tree, Anacardiaceae, Spain, s
6 *Acer pseudoplatanus*, Sycamore, Aceraceae, UK, s
7 *Tilia* sp., Lime, Tiliaceae, UK, s
8 *Heliocarpus americana*, Tiliaceae, Venezuela, s
9 *Tamarix pentandra*, Salt Cedar, Tamaricaceae, USA, s
10 *Anthiscus sylvestris*, Cow Parsley, Umbelliferae, UK, s
11 *Miconia* (*Tamonea*) sp., Melastomaceae, Venezuela, s
12 *Carica papaya*, Pawpaw, Caricaceae, India, s
13 *Eucalyptus* sp., Myrtaceae, Portugal, s
14 *Callistemon citrinus*, Bottle Brush, Myrtaceae, USA, s
15 *Lindenbergia indica*, Scrophulariaceae, India, s
16 *Calluna vulgaris*, Ling (Heather), Ericaceae, UK, s
17 *Ligustrum ovalifolium*, Privet, Oleaceae, Spain, s
18 *Olea europaea*, Olive, Oleaceae, Spain, s
19 *Fraxinus angustifolia*, Narrow Leaved Ash, Oleaceae, Spain, s
20 *Rungia pectinata*, Acanthaceae, India, s
21 *Plantago lanceolata*, Ribwort, Plantaginaceae, UK, s
22 *Plantago coronopus*, Buck's-horn Plantain, Plantaginaceae, Spain, s
23 *Sambucus nigra*, Elder, Caprifoliaceae, UK, s
24 *Senecio vulgaris*, Groundsel, Compositae, UK, s
25 *Parthenium hysterophorus*, Compositae, India s
26 *Solidago* sp., Golden Rod, Compositae, UK, s
27 *Artemisia vulgaris*, Mugwort, Compositae, UK, s
28 *Ambrosia deltoidea*, Triangle Leaf Bursage, Compositae, USA, s
29 *Taraxacum* sp., Dandelion, Compositae, UK, s

Plate 7
Coniferous pollen and fern spores

10μm

102 THE AIR SPORA

Plate 7. Coniferous pollen and fern spores

01 *Pinus sylvestris*, Scots pine, Pinaceae, UK, s
02 *Cedrus libani,* Cedar of Lebanon, Pinaceae, UK, s
03 *Podocarpus neriifolia*, Podocarpaceae, Singapore, s
04 *Taxus baccata*, Yew, Taxaceae, UK, s
05 *Juniperus communis*, Juniper, Cupressaceae, Sweden, s
06 *Pteridium aquilinum*, Bracken, fern spore, UK, s
07 fern spore, rain forest, Australia, t
08 *Asplenium nidus*, fern spore, Singapore, s
09 *Phyllitis scolopendrium*, Heart's Tongue, fern spore, UK, s
10 *Dicranopteris linearis*, fern spore, Singapore, s
11 *Stenochlaena palustris*, fern spore, Singapore, s
12 *Equisetum* sp., horsetail spore with elaters, UK, s
13 *Selaginella pulcherrina*, clubmoss spore, UK, s
14 *Lycopodium* sp., clubmoss spore, UK, s

Plate 8
Fungal spores –
Ascospores and
Uridinales

10μm

Plate 8. Fungal spores – Ascospores and Uridinales

1. *Didymosphaeria donacina*, ascospore, Singapore, c
2. *Ophiobolus graminis*, ascospore, [take-all of wheat] UK, s
3. *Leptosphaeria maculans*, ascospore, (see *Phoma lingam*) [canker of brassicas], UK, s
4. *Pringsheimia* type, Costa Rica, t
5. *Mycosphaerella capsellae* (see *Pseudocosporella capsellae*), [white leaf spot of brassicas] ascospore, UK, s
6. *Phaeosphaeria fuchlii*, ascospore, UK, t
7. *Phaeosphaeria nigrans*, ascospore, UK, t
8. *Micronectriella nivalis* (*Spaerulina*), ascospore, UK, t
9. *Pleospora herbarum*, ascospore, UK, s
10. *Pleospora* type, ascospore, UK, t
11. *Sporormia* type, ascospore, India, t
12. *Didymella* type, ascospore (hyaline), [late summer asthma in man], UK, t
13. *Didymella* type, ascospore (rough), UK, t
14. *Nectria cinnabarina*, [coral spot of woody plants] ascospore, UK, s
15. *Claviceps purpurea*, ascospore, [ergot], UK, s
16. *Venturia inaequalis*, ascospore, (see *Spilocaea pomi*) [apple scab], UK, s
17. *Passeriniella* type, ascospore, Costa Rica, t
18. *Neobulgaria pura*, ascospore, UK, s
19. *Sclerotinia sclerotiorum*, ascospores, [stem rot of many plants], UK, s
20. *Pyrenopeziza brassicae*, ascospores, (see *Cylindrosporium concentricum*) [light leaf spot of brassicas], UK, s
21. *Xanthoria parietina*, ascospore from lichen, UK, s
22. *Helvella crispa*, ascospore, UK, s
23. *Humaria granulata*, ascospore, UK, s
24. *Pyronema confluens*, ascospore, UK, s
25. *Melanospora zamiae*, ascospore, UK, s
26. *Chaetomium globosum*, ascospore, UK, c
27. *Chaetomium indicum*, ascospore, UK, c
28. *Sordaria fumicola*, ascospore, UK, c
29. *Daldinia concentrica*, ascospore, UK, s
30. *Xylaria polymorpha*, ascospore, UK, s
31. *Hypoxylon coccinium*, ascospore, UK, s
32. *Hypoxylon multiforme*, ascospore, UK, s
33. *Rosellinia aquila*, ascospore, UK, s
34. ascospore, UK, t
35. ascospore, Australia, t
36. ascospore, India, t
37. ascospore, India, t
38. ascospore, Costa Rica, t
39. ascospore, Costa Rica. t
40. ascospore, Costa Rica, t
41. ascospore, India, t
42. ascospore, Costa Rica, t
43. ascospore, Costa Rica, t
44. ascospore, Australia, t
45. ascospore, Costa Rica, t
46. ascospore, UK, t
47. *Puccinia graminis*, (a) teleutospore, (b) uredospore, [stem rust of cereals], UK, s
48. *Puccinia striiformis*, teleutospore, [yellow rust of wheat], India, s
49. *Puccinia striiformis*, uredospore, [yellow rust of wheat], UK, t
50. *Puccinia arachidis*, uredospore, [rust of peanuts], India, s
51. *Uromyces fabae*, uredospore, [chocolate spot of beans], UK, s
52. *Triphragmium ulmariae*, uredospore, UK, s
53. *Caeoma euphorbiae-geniculate*, aecidiospore, India, s
54. *Melampsoridiun betulinum*, uredospore, [rust of birch], UK, s
55. *Aecidium mori*, aecidiospore, [mulberry rust], India, s

IDENTIFICATION

Plate 9
Fungal spores –
Basidiospores and others

10 μm

106 THE AIR SPORA

Plate 9. Fungal spores - Basidiospores and others

1. *Agaricus bispora*, Cultivated Mushroom, basidiospore, UK, s.
2. *Lepiota racodes*, basidiospore, UK, s
3. *Amanita muscaria*, Fly Agaric, basidiospore, UK, s
4. *Amanita rubescens*, The Blusher, basidiospore, UK, s
5. *Bolbitius vitellinus*, basidiospore, UK, s
6. *Coprinus atramentarius*, Common Ink Cap, basidiospores, (a) profile, (b) face view, UK, s
7. *Coprinus micaceus*, Glistening Ink cap, basidiospore, UK, s
8. *Lacrymaria velutina*, Weeping Widow, basidiospore, UK, s
9. *Entoloma rhodopolium*, basidiospore, UK, s
10. *Nolanea staurospora*, basidiospore, UK, s
11. *Pluteus cervinus*, basidiospore, UK, s
12. *Hypholoma fascicularia*, Sulphur Tuft, basidiospore, UK, s
13. *Hypholoma hydophilum*, basidiospore, UK, s
14. *Pholiota squarrosa*, Shaggy Pholiota, basidiospore, UK, s
15. *Panaeolus (Psilocybe) foenisecii*, basidiospore, UK, s
16. *Panaeolus sphinctrinus*, basidiospore, UK, s
17. *Stropharia aeruginosa*, Verdigris Agaric, basidiospore, UK, s
18. *Armillaria mellea*, Honey Fungus, basidiospore, [white rot and death of trees], UK, s
19. *Collybia confluens*, Clustered Tough-shank, basidiospore, UK, s
20. *Collybia maculata*, Spotted Tough-shank, basidiospore, UK, s
21. *Laccaria amethystine*, Amethist Deceiver, basidiospore, UK, s
22. *Mycena crocata*, basidiospore, UK, s
23. *Mycena inclanata*, basidiospore, UK, s
24. *Tricholoma nudum*, Wood Blewit, basidiospore, UK, s
25. *Tricholoma album*, basidiospore, UK, s
26. *Boletus chrysenteron*, Red-cracked Boletus, basidiospore, UK, s
27. *Boletus elegans*, basidiospore, UK, s
28. *Boletus scaber*, basidiospore, UK, s
29. *Serpula (Merulius) lacrymans*, [dry-rot], basidiospore, UK, s
30. *Craterellus cornucopioides*, Horn of Plenty, basidiospore, UK, s
31. *Cortinarius elatior*, basidiospore, UK, s
32. *Gymnopilus penitrans*, basidiospore. UK, s
33. *Gymnopilus junonius (Pholiota penitrans)*, basidiospore. UK, s
34. *Inocybe geophylla*, basidiospore, UK, s
35. *Crepidotus mollis*, basidiospore, UK, s
36. *Tubaria furfuracea*, basidiospore, UK, s
37. *Fistulina hepatica*, Beefsteak Fungus, basidiospore, UK, s
38. *Ganoderma applanatum*, basidiospore, [wood decay], UK, s
39. *Bovista plumbea*, basidiospore, UK, s
40. *Lycoperdon giganteum (Calvatia gigantia)*, Giant Puff-ball, basidiospore, UK, s
41. *Lycoperdon perlatum*, basidiospore, UK, s
42. *Phallus impudicans*, Stink Horn, basidiospore, UK, s
43. *Heterobasidion annosum (Fomes)*, [conifer polypore root and butt rot], basidiospore, UK, s
44. *Pleurotus ostreatus*, Oyster Mushroom, basidiospore. UK, s
45. *Russula ochroleuca*, Common Yellow Russula, basidiospore, UK, s
46. *Russula vesca*, Bare-Toothed Russula, basidiospore, UK, s
47. *Lactarius blennius*, Slimy Milk-cap, basidiospore, UK, s
48. *Lactarius rufus*, Rufus Milk-cap, basidiospore, UK, s
49. *Chondrostereum (Stereum) purpureum*, [silver leaf of trees], basidiospore, UK, s
50. *Thelephora terrestris*, Earth Fan, basidiospore, UK, s

Plate 9. Fungal spores - Basidiospores and others *(continued)*

51 *Tilletia holci*, smut spore, UK, s
52 *Tilletia barclayana*, smut spore, [rice smut], India, s
53 *Tilletia caries*, smut spore, [wheat bunt], UK, s
54 *Tilletia* type, smut spore, India, t
55 *Urocystis agropyri*, smut spore, UK, s
56 *Sphacelotheca cruenta*, smut spore, [loose smut of sorghum], India, s
57 *Sphacelotheca reiliana*, smut spore, [head smut of sorghum], India, s
58 *Sphacelotheca sorghi*, smut spore, [covered smut of sorghum], India, s
59 *Tolyposporium ehrenbergii*, smut spore, [long smut of sorghum], India, s
60 *Ustilago avenae*, smut spore, [oat smut], UK, s
61 *Enteridium (Reticularia) tycoperdon*, myxomycete spore, UK, s
62 *Badhamia utricularis*, myxomycete spore, UK, s
63 *Leocarpus fragilis*, myxomycete spore, UK, s
64 *Fuligo septica*, myxomycete spore, UK, s
65 myxomycete/smut type spore, India, t
66 myxomycete/smut type spore, India, t
67 myxomycete/smut type spore, India, t
68 *Peronospora parasitica*, sporangium, [downy mildew of brassicas], UK, s
69 *Phytophthora infestans*, sporangium, [potato blight], UK, s
70 *Albugo* sp., conidium, [white rust], UK, t
71 *Peronosclerospora sorghi*, a) oospore, b) conidium, [downy mildew of sorghum], India, s
72 *Conidiobolus obscurus*, primary conidium, UK, s
73 *Entomophthora planchoniana*, primary conidium, UK, s
74 *Erynia neoaphidis*, primary conidium UK, s
75 *Neozygites freserii*, primary conidium UK, s
76 *Absidia corymbifera*, sporangiospores, [bovine mycotic abortion] UK, c
77 *Absidia ramosa*, sporangiospores, UK, c
78 *Mucor spinosus*, sporangiospores, UK, c
79 *Rhizopus nigricans*, sporangiospores, UK, c

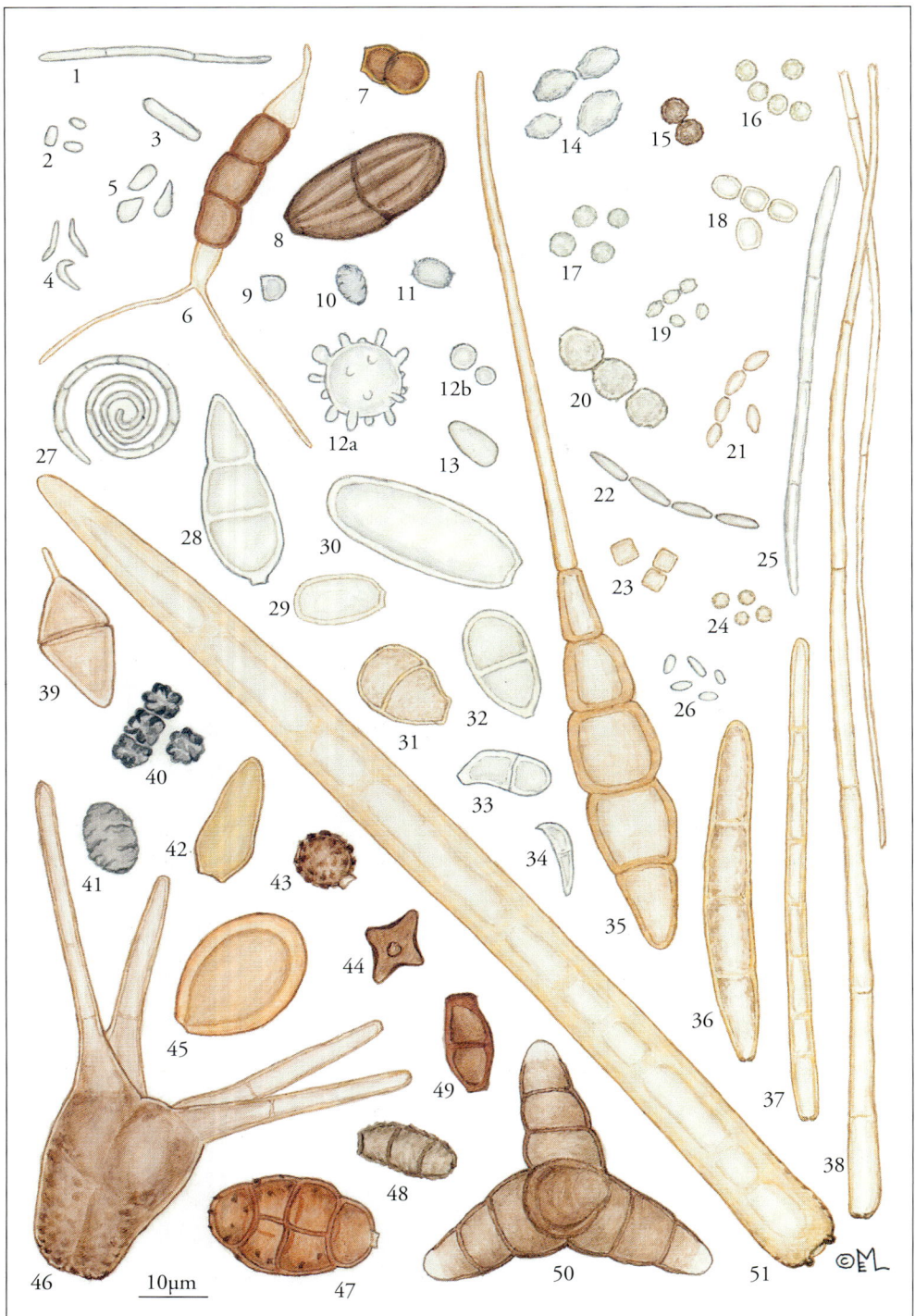

Plate 10
Fungal spores -
Mitosporic

Plate 10. Fungal spores – Mitosporic

01 *Septoria tritici*, conidium, [wheat leaf spot], UK, s
02 *Phoma lingam*, conidia. (see *Leptosphaeria maculans*) [canker of brassicas], UK, s
03 *Cylindrosporium concentricum*, conidium (see *Pyrenopeziza brassicae*), [light leaf spot of brassicas], UK, s
04 *Tilletiopsis* sp., ballistosporres, UK, t
05 *Sporobolomyces* sp., ballistosporres, UK, c
06 *Pestalotiopsis theae*, conidium, [grey blight of tea], India, c
07 *Botryodiplodia* sp. conidium, India, s
08 *Botryodiplodia acerina*, conidium, India, s
09 *Blastomyces dermatitidis*, conidium, [blastomycosis in man and animals], USA, c
10 *Chrysosporium* sp., conidium, UK, c
11 *Coccidioides immitis*, conidium, [coccidiomycosis of man and animals], UK, c
12 *Histoplasma capsulatum*, a) macroconidium, b) microconidium, [histoplasmosis in man and animals], UK, c
13 *Sporotrichum* sp., conidium, UK, c
14 *Aspergillus glaucus* (series), conidium, UK, c
15 *Aspergillus niger*, conidia, UK, c
16 *Aspergillus fumigatus*, conidia, [asthma, allergic alveolitis and aspergillosis of man], UK, c
17 *Penicillium frequentens*, conidia, [suberosis in man], UK, c
18 *Penicillium chrysogenum* conidia UK, c
19 *Penicillium marneffei*, conidia, UK, c
20 *Scopulariopsis brevicaulis*, conidia, UK, c
21 *Paecilomyces varioti*, conidia, UK, c
22 *Paecilomyces* type, conidia, Costa Rica, t
23 *Wallemia sebi*, conidia, UK, s
24 *Trichoderma viride*, conidia, UK, c
25 *Pseudocercosporella capsellae*, (see *Mycosphaerella capsellae*) [white leaf spot of brassicas], conidium, UK, s
26 *Verticillium dahliae*, conidia, [wilt of plants], UK, s
27 *Helicomyces* type, conidium, UK, t
28 *Pyricularia oryzae*, conidium, [rice blast], India, s
29 *Botrytis* sp., conidium, UK, c
30 *Blumeria graminis*, (= *Oidium, Erysiphe*), conidium, [powdery mildew], UK, t
31 *Polythrincium trifolii*, conidium, UK, s
32 unknown conidium, Costa Rica, t
33 *Trichothecium roseum*, conidium, UK, c
34 *Rhynchosporium secalis*, conidium, [leaf blotch of barley and rye]. UK, s
35 *Trichoconis padwickii*, conidium, [stack burn and leaf spot of rice], India, t
36 *Cercosporidium personatum*, conidium, [late leaf spot of peanuts], India, s
37 *Cercospora arachidicola*, conidium, [early leaf spot of peanuts], India, s
38 *Cercospora* sp., conidium, India, t
39 *Beltrania* sp. conidium, India, t
40 *Memnoniella echinata*, conidia, India, s
41 *Stachybotris* sp., conidium, UK, c
42 *Spilocaea pomi*, conidium, (see *Venturia inaequalis*), [apple scab], UK, s
43 *Humicola lanuginosa*, conidium, UK, s
44 *Humicola stellata*, conidium, UK, c
45 *Acremoniella atra*, conidium, UK, c
46 *Tetraploa aristata*, conidium, India, t
47 *Pithomyces chartarum*, conidium, [facial eczema of sheep], UK, s
48 *Pithomyces maydicus*, conidium, Singapore, c
49 *Bispora monilioides*, conidium, UK, s
50 *Asterosporium* sp., conidium, UK, t
51 *Corynespora* sp., conidium, India, t

Plate 11
Fungal and moss spores, diatoms and algae

10 µm

Plate 11. Fungal and moss spores, diatoms and algae

1   *Cladosporium herbarum*, conidia, UK, t
2   *Cladosporium cladosporioides*, conidia, UK, s
3   *Alternaria brassicae*, conidium, [dark leaf spot of brassicas], UK, c
4   *Alternaria infectoria*, conidium, UK, c
5   *Alternaria brassicicola*, conidium, [leaf spot of brassicas], UK, c
6   *Alternaria alternata*, conidium, UK, c
7   *Helmimthosporium (Drechslera)* sp., conidium, UK, t
8   *Drechslera oryzae*, conidium, [brown spot of rice], India, s
9   *Torula herbarum*, conidium, UK, t
10 *Torula* sp., conidium, India, t
11 *Cryptostroma corticale*, conidium, [sooty bark of sycamore], UK, s
12 *Arthrinium (Papularia) arundinis*, conidia: (a) face view, (b) edge view, UK, c
13 *Epicoccum* sp., conidium, UK, t
14 *Epicoccum* sp., conidium, Costa Rica, t
15 *Oncopodiella* type, conidium, UK, t
16 *Curvularia lunata*, conidium, Singapore, c
17 *Curvularia lunata*, conidium, India, t
18 *Nigrospora* sp. conidium, India, c
19 *Ceratosporiella* type, conidium, India, t
20 *Stemphylium* sp., conidium, UK, t
21 *Exosporium* sp., conidium, UK, t
22 *Spegazzinia lobulata*, conidium, India, t
23 *Spegazzinia deightonii*, conidium, India, t
24 *Spegazzinia tessartha*, conidium, India, t
25 *Sphacelia sorghii*, conidium, [ergot of sorghum], India, s
26 *Tubercularia vulgaris*, conidium, (see *Nectria cinnabarina*), [coral spot], UK, s
27 *Fusarium graminiarum*, conidium, UK, s
28 *Schistidium antartica*, moss spore, Antarctica, s
29 *Funaria hygrometrica*, moss spore, UK, s
30 *Barbula fallax*, moss spore, UK, s
31 *Bartramina patens*, moss spore., Antarctica, s
32 *Bryum algens*, moss spore, Antarctica, s
33 *Ceratodon purpureus*, moss spore, UK, s
34 diatom, India, t
35 diatom, UK, t
36 *Cyclotella* type, diatom, India, t
37 *Gloeocapsa* sp, algal group, UK, t
38 *Cladonia* sp., lichen soredium, UK, s

IDENTIFICATION

Plate 12
Miscellaneous

10μm

Plate 12. Miscellaneous

01 skin scales with bacteria, UK, t
02 *Thermoactinomyces* sp., actinomycete spores, UK, c
03 *Sacharopolyspora rectevergilla*, actinomycete spores, [bagassosis in man], UK, c
04 *Streptomyces* sp., actinomycete spores UK, s
05 *Bacillus subtillus*, bacteria, UK, c
06 *Staphylococcus aureus*, bacteria, UK, c
07 plant hair, UK, t
08 cotton fibre, UK, s
09 hyphal fragment, UK, t
10 pro-xylem, UK, t
11 meta-xylem, UK, t
12 cat fur, UK, s
13 piece of feather, UK, s
14 *Tetramitus sp.*, amoeboid cyst, Tristan da Chuna, t
15 mite pellet, UK, t
16 part of mite leg, UK, t
17 insect scale, UK, t
18 moth scale, UK, s
19 a – d insect hairs, UK, t
20 part of insect leg, UK, t
21 part of insect compound eye, Australia, t
22 *Thecamoeba* sp., UK, t
23 diesel particles, UK, s
24 small talc (magnesium silicate) particles, UK, s
25 combustion product, UK, t
26 particles from soil, UK, t
27 fly ash spheres, UK, t
28 particles from soil, UK, t
29 carbon shards from bonfire, UK, t
30 tyre roll, UK, t
31 combustion product, UK, t
32 combustion product, UK, t

Appendix

1. Recipes

WARNING – Phenol is poisonous and can be absorbed through the skin. Plastic utensils must not be used with mixtures containing Phenol.

Hexane should only be used in a fume cupboard and using eye protection.

Petroleum jelly - to coat slides, rods, etc. for spore trapping

150 ml petroleum jelly (100 g jar, Vaseline)
18 g paraffin wax of low melting point
10 g phenol

Heat in water bath to melt, stir with glass rod to mix.
To use, dissolve in Hexane or melt in a water bath.

Gelvatol (Polyvinyl alcohol, Moviol) – permanent mountant

35 g Gelvatol
50 ml Glycerol
100 ml distilled water
2 g phenol

Put Gelvatol and phenol in water and leave to stand overnight.
Add the glycerol and heat in water bath or warm gradually in a microwave, stir to mix.

Lactophenol – temporary mountant

20 g phenol
16 ml lactic acid
31 ml glycerol
20 ml Distilled water

Put ingredients together and stir, warm gradually to dissolve the phenol.

Glycerine jelly – sticky surface and permanent mountant

10 g gelatine
54 ml glycerol
60 ml distilled water
1.4 g phenol

Put all ingredients together and heat in a water bath or warm gradually in a microwave, stir frequently till melted
Melt to use, **do not boil.**

Stains

Trypan Blue and **Cotton Blue** can be used to stain fungal spores.
Basic Fuchsin and **Safranin** can be used to stain pollen.

Make up a small amount of stain and add just enough to the mountant to get the required depth of stain.

2. Suppliers

Stains and other chemicals

BDH,
Mail Merek Ltd.,
Hunter Boulivard, Magna Park,
Lutterworth
Leics, LE17 4XN, UK

M.J. Patterson (Scientific) Ltd,
Bramingham Business Park,
Enterprise Way,
Luton,
Beds, LU5 4UB, UK

Burkard spore trap, Gelvatol, Melinex tape and accessories.

Burkard Manufactoring Co. Ltd.,
Woodcock Hill Industrial Estate,
Ricmansworth,
Herts, WD3 1PJ, UK

Burkard Scientific Ltds.,
PO Box 55, Uxbridge
Middx, UB8 2RT, UK

Motor for whirling arm trap

McLennon Servo Supplies
Unit L
Yorktown Industrial Estate
Doman Road, Camberly
Surrey, GU15 3DE, UK

Graticules and micrometers

Graticules Limited
Marley Road
Tonbridge
Kent, TW9 1RN, UK

Other Suppliers

Advances in Life Sciences
9 Market Place, Brackley
NN13 7AB, UK

Biotrace International plc
The Science Park
Bridgend
CF31 3NA, UK

Copley Scientific Limited,
Colwick Quays Business Park,
Private Road No. 2, Colwick,
Nottingham NG4 2JY, U.K.

Lanzoni s. r. l.,
Via Michelino, 93/B,
40127 BOLOGNA,
Italy

SKC Ltd.,
11 Sunrise Park,
Higher Shaftsbury Road,
Blandford Forum, Dorset
DT11 8ST, UK

VWR International
Lutterworth, Leicester
LE17 4XN, UK

3. Web Pages

CABI Bioscience and CABI Publishing
http://www.cabi.org

British Aerobiology Federation
http://pollenuk.worc.ac.uk/baf/BAF.html

National Pollen and Aerobiology Research Unit
http://pollenuk.worc.ac.uk/Next.htm

Midlands Asthma and Allergy Research Association (MAARA)
http://www.maara.org

International Association for Aerobiology (IAA)
http://www.isac.cnr.it/aerobio/iaa/index.html

Italian Association of Aerobiology
http://www.isao.bo.cnr.it/%7Eaerobio/aia/index.html

For more information
http:www.helios.bto.ed.ac.uk/bto/microbes/airborne.htm#Top

4. Counting sheets

Two examples of counting sheets for a Burkard trap and one for the whirling arm trap are given. Counting sheets can be varied according to requirement. Records can be recorded mechanically.

Counting sheet for recording two-hourly pollen or spore counts and combined for daily pollen counts using the 12 transverse traverses method

STATION _____   DATE _____

| Objective Magnification: _____ | Eyepiece Magnification: _____ | Traverse width (µm): _____ |

Traverse number and direction	Time (BST)	Vernier reading	N⁰· particles counted
1 ↑	10:00		
2 ↓	12:00		
3 ↑	14:00		
4 ↓	16:00		
5 ↑	18:00		
6 ↓	20:00		
7 ↑	22:00		
8 ↓	00:00		
9 ↑	02:00		
10 ↓	04:00		
11 ↑	06:00		
12 ↓	08:00		
Total number of particles counted in 24 h			
Multiplied by correction factor (CF) of:		Particles m^{-3} air	

Notes:

Counted by: _____

APPENDIX

Counting sheet for recording daily spore counts using a mean of two longitudinal traverses method over 14 days (normally data are imported into a computer spreadsheet to calculate spores m^{-3} automatically, using the appropriate correction factor

TRAP : _____   SPECIES : _____

| Objective Magnification: _____ |
| Traverse width (µm): _____ |

| Eyepiece Magnification: _____ |
| Correction factor: _____ |

Date	Slide N°	Count A(→)	Count B (←)	Mean	Spores m^{-3}

Counted by : _____

Counting sheet for recording spore counts from whirling arm traps using five transverse traverses per quarter arm-section

TRAP: _____ SPECIES: _____

| Objective Magnification: | Eyepiece Magnification: | Traverse width (μm): | Correction factor: |

Trap Nº	Date/Time	Count Arm 1					Count Arm 2					Total	Spores per arm	Spores m^{-3}

Glossary

abiotic particles - particles that are not of biological origin e.g. soot or clay.
actinomycetes - filamentous gram positive bacteria
aerobiology - the study of biological particles present in air, their occurrence, dispersal and impact.
aerodynamic diameter – diameter of a spherical particle with an equivalent terminal velocity or fall speed.
aerosol - material finely divided and suspended in air or other gaseous environments.
air spora - the population of biological particles present in the air.
alga – (pl. algae) aquatic lower plants containing photosynthetic agents.
allergen – a substance capable of causing an allergic reaction.
allergy - a reaction of the body's immune system to the presence of a foreign substance e.g.; hay fever or asthma.

anamorph – see imperfect stage.
anemophilous plants - wind pollinated.
angiosperms – seed bearing plants with ovules protected in the ovary, see: monocotyledons and dicotyledons.
apothecia – cup shaped structures containing asci.
ascomycetes – large group of fungi with spores formed in asci.
ascospore - spore produced in an ascus.
ascus – (pl. asci) sac-like cell in which ascospores (generally eight) are produced.
atmosphere - layer of gas surrounding the earth.

bacteria – microscopic organisms with genetic material not bounded by a nuclear membrane.
basidiomycetes – large group of fungi with spores (usually four) produced from a basidium.
basidiospore - spore produced from a basidium.
bioaerosol - an aerosol of particles of biological origin or activity suspended in air. Particle size may range from aerodynamic diameters of ca. 0.5 to 100 μm.
boundary layer - lowest layers of air near the earth surface.
BST – British Summer Time.
Burkard trap – Seven-day recording volumetric spore trap used by many for the daily spore count.

calibration – determining the accurate measurement of a device.
cascade impactor - a four-stage volumetric spore trap
concentration - number of items per given unit of media e.g. number of particles m^{-3} of air.
conidium – (pl. conidia) asexually produced fungal spores.
conifers – see: gymnosperms.
culture – organism grown on culture media.
cyclone - 1: a centrifugal device to collect particles from air. 2: low pressure weather system with air circulation in a clockwise direction.

daily counts – daily counts of pollen or spores, often for allergy sufferers.
daily (diurnal) periodicity – the cycle of high and low production of pollen and spores during a day.
deposition – accumulation of particles on a surface.
diatom – microscopic alga having a intricately sculptured siliceous cell wall.
dicotyledons – flowering plants with two embryonic seed leaves.
diffusion – dispersal of molecules or particles into a medium.
dispersal – moving of particles over a wide area.

eddy - circulating currant of air.
ELISA – Enzyme-Linked Immunosorbent Assay: an immunological diagnostic technique
entomophilous plants – insect pollinated plants.

fall speed – see: terminal velocity.
ferns – spore bearing vascular plants, belong to the Pteridophyta.
filter – material used to trap particles present in a fluid.
filtration – process of filtering particles from a fluid.
flowmeter – an instrument to measure the volume of air passing through a sampler.

GMT – Greenwich Mean Time
gradient – change with space or time in the number or frequency of a feature e.g. high to low.
graticule – a scale on glass in the eyepiece of a microscope to enable measurement of objects.
gust – temporary increase in wind speed significantly greater than the mean wind speed.
gymnosperms – seed bearing plants with unprotected ovaries: e.g. conifers.

Hirst spore trap – the first automatic volumetric spore trap.
hourly concentration – concentration of particles in the air for each hour.
hyaline - descriptive term for a structure that is transparent or nearly so.

immunoassay – test for presence of an allergen e.g.: a specific organism or protein (see: ELISA).

immunology – study of the immune system and applications such as using antibodies specific to an antigen for the detection or identification of the organism that produces the antigen
impact – 1: have a strong effect as in the aerobiology pathway. 2: deposited by collision.
impaction – the sticking of airborne particles onto a surface following an active collision.
impactor – an apparatus for catching particles by impaction on a sticky surface or in a liquid.
imperfect stage - part of the lifecycle of certain fungi in which asexual reproduction takes place.
impinger - device to trap particles by their impaction onto a sticky surface or capture into liquid.

laminar boundary layer – microscopically thin layer of still air next to object surfaces, see: boundary layer,
liberation – release or setting a particle free.
lichen – symbiotic association of an alga and a fungus.

macrofungi – fungi having large sporocarps e.g. toadstools.
mesophilic - organisms with optimum growth temperature of 20-35º C.
micrometer – an instrument for measuring very small distances.
microorganism (microbe) - a microscopic organism.
mitosporic – imperfect fungi, spores produced asexually.
monocotyledon – flowering plant with a single embryonic leaf.
mosses – green plants with simple leaves and stems but no roots, belong to the Bryophyta.
mountant - substance used to embed an object between a microscope slide and coverslip.
myxomycete - slime mould.

nematode - unsegmented worm that can be saprophytic or parasitic on plants or animals
number of biological particles per m³ air – standard reference for quantifying the number of airborne biological particles, see: concentration.

orifice - the opening through which air is drawn into a sampling device.

pathogenic - able to cause disease.
perfect stage - part of the lifecycle of a fungus that undergoes sexual reproduction.
PCR, Polymerase Chain Reaction - a method to replicate a specific section of DNA
pollen - spore-like male gamete of a higher plant that fertilises an ovum leading to the production of a seed, may act as an allergen causing hay fever.
protist – single celled organisms e.g. protozoa.

relative humidity – measure of water vapour in air relative to the total amount possible to be present at that temperature.
release - see: liberation.

sampling – to take a representative sample of a medium, such as air, to determine its content.
sedimentation – accumulation of particles on a surface by settling.
spore - microscopic reproductive organ of a bacterium, fungus, moss or fern.
stage micrometer – a microscope slide with a scale 1 mm in length and divided into 100 divisions.
stroboscope – instrument producing a very bright flashing light used to measure rotation speeds.
suction trap – a trap which samples air by drawing a known volume of air into it under suction.

teleomorph – see: perfect stage.
teleutospore - a thick-walled resting spore of the rust fungi (and also used for the smut fungi although ustilospore is more correct) that produces basidiospores by sexual reproduction
terminal velocity – see: fall speed, the maximum speed a particle reaches when falling through still air due to a balance of acceleration due to gravity and drag due to air resistance.
thermophilic – organisms with optimum growth temperature of 40-50º C.
trace - the deposit of particles on a sticky surface.
traverse - the area counted along or across the trace.

uredospore - asexually produced spore of rust fungi with many successive cycles of infection.
Uridinales – rust fungi.

vernier scale – movable scale on a microscope stage to enable the position on a microscope slide to be noted.
viability – measure of the ability to live and grow.
virus – sub-microscopic organism comprising a strand of nucleic acid and a protein coat.
volumetric spore trap – type of trap with which the volume of air sampled per unit time is known.

wind – current of air.
wind tunnel – a tunnel through which air is drawn to enable the study of airborne particle release, dispersal and deposition.

References

Agranovski, I.E., Agranovski, V., Reponen, T., Willeke, K. and Grinshpun, S.A. 2002, Development and evaluation of a new personal sampler for culturable airborne microorganisms, *Atmospheric Environment*, **36**:889-898.

Ainsworth G.C, 1961, *Ainsworth and Bisby's Dictionary of the Fungi*, 5th ed., CMI, Kew, pp 547.

Allitt, U., 1979, The visual identification of airborne fungal spores, *Proceedings of the 1st International Conference on Aerobiology, August 1978, Berichte* 5/79, 139-143, Eric Schmidt Verlag, Berlin.

Andersen, A.A., 1958, New sampler for the collection, sizing and enumeration of viable airborne particles, *Journal of Bacteriology*, **76**:471-484.

Avila, R. and Lacey, J. 1974, The role of *Pennicillium frequentens* in suberosis (respiratory disease in the cork industry), *Clinical Allergy*, **4**:109-117.

Aylor, D.E., McCartney, H.A. and Bainbridge, A., 1981, Deposition of particles liberated in gusts of wind, *J. Appl. Meteorol.*, **20**:1212-1221.

BAF, 1995, *Airborne Pollens and Spores: A guide to trapping and counting*, The British Aerobiology Federation, Harpenden, pp 59.

Bainbridge, A. and Brent, K.J., 1999, John Malcolm Hirst 20 April 1921 – 30 December 1997, *Biog. Mems. Fell. R. Soc. Lond*, **45**:219-238.

Bargagli, R., Broady, P.A. and Walton, D.W.H., 1996. Preliminary investigation of the thermal biosystem of Mount Rittmann fumaroles (northern Victoria Land, Antarctica). *Antarctic Science*, **8**:121-126.

Barnet, H.L. and Hunter, B.B., 1998, *Illustrated Genera of Imperfect Fungi*, 4th ed., The American Phytopathological Society, St Paul.

Blackley, C.H.,1873, Experimental researches on the causes and nature of Catarrhus Aestivus (Hay Fever or Hay Asthma). Baillière, Tindall and Cox, London, (Reprinted: Dawson, London, 1959)

Boros, Á., Járai-Komlódi, M., Tóth, Z. and Nilsson, S., 1993, *An Atlas of Recent European Bryophyte Spores*, Scientia Publishing, Budapest, pp 321.

Bovallius, Å., Bucht, B., Roffey, R. and Anas, P., 1978, Long-Range Air Transmission of Bacteria. *Applied and Environmental Microbiology*, **35**:1231-1232.

Brown, J.K.M. and Hovmøller, M.S., 2002, Epidemiology - Aerial dispersal of pathogens on the global and continental scales and its impact on plant disease, *Science*, **297**:537-541.

Buller, A.H.R., (1915), Micheli and the discovery of reproduction in fungi, *Trans. R. Soc. Can.* (Ser.3), **9**, Sect. IV, 1-25.

Bulloch, W., (1938), *The History of Bacteriology*, Clarendon Press, Oxford, pp 422.

Burr, M.L., Emberlin, J., Treu, R., Cheng, S., Pearce, N.E., and the ISAAC Phase 1 Study Group, 2003, Pollen counts in relation to the prevalence of allergic rhinoconjunctivitis, asthma and atopic eczema in the international study of asthma and allergies in children (ISAAC), *Clin. Exp. Allergy*, **33**:1675-1680.

Burt, P.J.A., Rosenberg, L. J., Rutter, J., Ramirez, F. and Gonzales, H., 1999, Forecasting the airborne spread of *Micosphaerella fijiensis*, a cause of black Sigatoka disease on bananas: estimations of numbers of perithecia and ascospores to aid forecasting, *Annals of Applied Biology*, **135**:367-377.

CABI Biosciences (previously CMI and IMI), 1964 onwards, *Descriptions of Fungi and Bacteria*, CAB International, Wallingford.

Cadham, F.T., 1924, Asthma due to grain rusts, *J. Am. Med. Ass.* **83**:27

Calderon, C., Ward, E., Freeman, J. and McCartney, H.A., 2002, Detection of airborne fungal spores sampled by rotating-arm and Hirst-type spore traps using polymerase chain reaction assays, *Journal of Aerosol Science,* **33**:283-296.

Campbell, C.K., Johnson, E.M., Philpot, C.M. and Warnock, D.W., 1996, *Identification of Pathogenic Fungi*, London Public Health Laboratory Service, pp 298.

Carmichael, J.W., Kendrick, W.B., Conners, I.L. and Sigler, L., 1980, *Genera of Hyphomycetes*, Univ. of Alberta Press, Edmonton, pp 386.

Caulton, E., 1988, Rabbit faeces as a source of pollen for dietary and habitat studies – a suggestion for project work suitable for sixth form and higher students, *Journal of Biological Education*, **22**:37-40.

Caulton, E., Carmichael, R., Minebois, E. and Vaschetti, C., 1997, Calendar of Allergenic Pollens in Scotland, *Scottish Centre for Pollen Studies*, Napier University, Edinburgh.

Caulton, E., Keddie, S., Carmichael, R. and Sales, J., 1999, A ten year study of the incidence of the spores of bracken (*Pteridium aquilinum*) in an urban rooftop airstream in South-east Scotland, Abstract in *Bracken Fern: Toxicity, biology and Control, Proceedings of the International Bracken Group Conference, Manchester, 1999*, Taylor, J.A. and Smith, R.T., eds.

Chalmers, M.O., Harper, M.A. and Marshall, W.A., 1996, *An Illustrated Catalogue of Airborne Microbiota from the Maritime Antarctic,* British Antarctic Survey, Cambridge, pp 175.

Chamberlain, A.C., 1967, Cross-Pollination between Fields of Sugar Beet. *Quarterly Journal of the Royal Meteorological Society*, **93**:509-513.

Chamberlain, A.C., 1975, The movement of particles in plant communities, in *Vegetation and the atmosphere*, J.L. Monteith, ed., Academic Press, New York, pp 155-203.

Chen, H., Deng, G., Li, Z., Tian, G., Li, Y., Jiao, P., Zhang, L., Liu, Z., Webster, R.G., Yu, K., 2004,The evolution of H5N1 influenza viruses in ducks in southern China, *Proc Natl. Acad. Sci. USA*, **101**:10452-10457.

Cheng, Y.S., Barr, E.B., Fan, B.J., Hargis, P.J., Rader, D.J., O'Hern, T.J., Torczynski, J.R., Tisone, G.C., Preppernau, B.L., Young, S.A., and Radloff, R.J., 1999, Detection of bioaerosols using multiwavelength UV fluorescence spectroscopy, *Aerosol Science and Technology* **30**:186-201.

Convey, P., Smith, R.I.L., Hodgson, D.A., Peat, H.J., 2000. The flora of the South Sandwich Islands, with particular reference to the influence of geothermal heating. *Journal of Biogeography* **27**:1279-1295.

Corden, J.M., Millington, W.M., 2001, The long-term trends and seasonal variation of the aeroallergen *Alternaria* in Derby, UK, *Aerobiologia*, **17**:127-136.

Corden, J.M., Millington, W.M., 1994, *Didymella* ascospores in Derby, *Grana,* **33**:104-107.

Corden, J.M., Millington, W.M., Bailey J., Brookes M., Caulton E., Emberlin J., Mullins J., Simpson C. and Wood A., 2000, UK regional variations in *Betula* pollen (1993-1997), *Aerobiologia* **16**:227-232.

Corden, J.M., Millington, W.M. and Mullins J., 2003, Long term trends and regional variation in the aeroallergen *Alternaria* in Cardiff and Derby UK – are differences in climate and cereal production having an effect, *Aerobiologia***, 19**:191-199.

Cox, C.S., 1987, *The Aerobiological Pathway of Microorganisms*, John Wiley, Chichester.

Cox, C.S. and Wathers, C.M. eds. 1995, *Bioaerosols Handbook*, Lewis Publishers, Boca Raton, pp 623.

Crook, B., 1995a, Inertial samplers: biological perspectives, in: *Bioaerosols Handbook*, Cox, C.S. and Wathers, C.M., eds. Lewis Publishers, Boca Raton, pp 247-267.

Crook, B., 1995b, Non-inertial samplers: biological perspectives, in: *Bioaerosols Handbook*, Cox, C.S. and Wathers, C.M., eds., Lewis Publishers, Boca Raton, pp 269-283.

Crook, B. and Lacey, J., 1988, Enumeration of airborne micro-organisms in work environments, in: *Indoor and Ambient Air Quality*, R. Perry and P.W. Kirk, eds. Selper Ltd, London.

Crook, B. and Lacey, J., 1991, Airborne allergenic microorganisms associated with mushroom cultivation, *Grana*, **30**:446-449.

Crook, B. and Swan, J.R.M., 2001, Bacteria and other bioaerosols in industrial workplaces, in: *Microorganisms in Home and Indoor Work Environments*, Flannigan, B., Samson, R.A. and Miller, J.D.,eds., Taylor and Francis, London, pp 69-82.

Cunningham, D.D., 1873, *Microscopic Examinations of Air*, Government Printer, Calcutta, 58pp.

D'Amato, G., Spieksma, F.Th.M. and Bonina, S., 1991, Allergenic pollen and pollinosis in Europe, Blakwell, Oxford.

Darke, C.S., Knowelden, J., Lacey, J. and Milford Ward, A., 1976, Respiratory disease of workers harvesting grain, *Thorax*. **31**:294-302.

Darwin, C., (1846), An account of the fine dust which often falls on vessels in the Atlantic Ocean, *Q. Jl geol. Soc. Lond.*, **2**:26-30.

Day, J.P., Kell, D.B., and Griffith, G.W., 2002, Differentiation of *Phytophthora infestans* sporangia from other airborne biological particles by flow Cytometry, *Applied and Environmental Microbiology* **68**:37-45.

Dixon, P.M., McGorum, B. and Caulton, E., 1992, A study of pasture associated respiratory disease in horses, Index of equine research in the British Isles and Ireland, *British Equine Veterinary Association Trust*, **8**:22

Dobell, C., 1932, *Antony van Leewenhoek and his 'Little Animals'*, Bale & Danielsson, London, pp 435

Donaldson, A.I., Alexandersen, S., 2002, Predicting the spread of foot and mouth disease by airborne virus, *Rev. Sci. Tech.*, **21**:569-575.

Edmonds, R.L., 1972, Collection Efficiency of Rotorod Samplers for Sampling Fungus Spores in Atmosphere, *Plant Disease Reporter*, **56**:704-708.

Edmonds, R.L., ed., 1979, Introduction, in: *Aerobiology: The Ecological Systems Approach*, US/IBP Synthesis Series **10**. Dowden, Hutchinson & Ross, Inc.: Stroudsberge, PA.

Edmonds, R.L. and Benninghoff, W.S., 1973, *Aerobiology and its Modern Applications,* Report No. 3, Aerobiology Component, U.S. Component of the International Biological Program.

Emberlin, J. 1997, Grass, tree and weed pollens, in: *Allergy and Allergic Diseases*, Vol. 2, A.B. Kay, ed., Blackwell Science Ltd, Oxford.

Emberlin, J. and Baboonian, C., 1995, The development of a new method of sampling air-borne particles for immunological analysis. In *XVI European Congress of Allerology and Clinical Immunology* (Edited by Basomba, A., Hernandez, M. D. and de Rojas, F.), pp 39-43, Monduzzi Editore, Bologna, Italy.

Emberlin, J., Jones, S., Bailey, J., Caulton, E., Corden, J., Dubbels, S., Evans, J., McDonagh, N., Millington, W., Mullins, J., Russell, R. and Spencer, T., 1994, Variation in the start of the grass pollen season at selected sites in the United Kingdom 1987 – 1992, *Grana*, **33**:94-99.

Evans, I.A., 1987, Bracken carcinogenicity, in: Reviews of Environmental Health, G.V. James, ed., *Int. Q. Sci. Rev.* **7**:169-199.

Eversole, J.D., Cary, W.K., Scotto, C.S., Pierson, R., Spence, M., and Campillo, A.J., 2001, Continuous bioaerosol monitoring using UV excitation fluorescence: Outdoor test results, *Field Analytical Chemistry and Technology* **5**:205-212.

Feinberg, S.M. 1935, Mould allergy: its importance in asthma and hay fever, *Wisconsin med. J.* **34**:254

Fitt, B.D.L., and Bainbridge, A., 1983, Dispersal of *Pseudocercosporella herpotrichoides* spores from infected wheat straw, *Phytopath Z.*, **106**:214-225.

Fitt, B.D.L. and McCartney, H.A., 1986, Spore dispersal in relation to epidemic models, in: *Plant Disease Epidemiology*, Vol. 1, Leonard, K.J. and Fry, W.E. eds., Macmillan Publishing Co., pp 311-346.

Fitt, B.D.L., McCartney, H.A. and Walklate, P.J., 1989, The role of rain in dispersal of pathogen inoculum, *Ann. Rev. Phytopathol.*, **27**:241-270.

Fitt, B.D.L., Walklate, P.J., McCartney, H.A., Bainbridge, A., Creighton, N.F., Hirst, J.M., Lacey, M.E. and Legg, B.J., 1986, A rain tower and wind tunnel for studying the dispersal of plant pathogens by rain and wind, *Ann. appl. Biol*, **109**:661-671.

Flannigan, B. 2001, Microorganisms in indoor air, in: *Microorganisms in Home and Indoor Work Environments*, Flannigan, B., Samson, R.A. and Miller, J.D. eds., Taylor and Francis, London, pp 17-31.

Flannigan, B., Samson, R.A. and Miller, J.D. eds., 2001, *Microorganisms in Home and Indoor Work Environments*, Taylor and Francis, London, pp 490.

Fraaije B.A., Cools H.J., Fountaine J., Lovell D.J., Motteram J., West J.S. and Lucas J.A. (2005) QoI resistant isolates of *Mycosphaerella graminicola* and the role of ascospores in further spread of resistant alleles in field populations. *Phytopathology* 95: 933-941.

Frankland, A.W. and Gregory, P.H., 1973, Allergenic and agricultural implications of airborne ascospores from a fungus *Didymalla exitalis*, *Nature*, **245**: 336-337.

Frankland, P.F., 1886, The distribution of microorganisms in air, *Proc. R.. Soc.*, **40**:506-526.

Frankland, P.F. and Hart, T.G., 1887, Further experiments on the distribution of microorganisms in air (by Hesse's method), *Proc. R. Soc.*, **42**:267-282.

Fraser, M.A., Caulton, E. and McNeil, P.D., 2001, Examination of faecal samples as a method of identifying pollen exposure in the dog, *Verterinary Record*, **149**:424-426.

Free, J. B., Williams, I. H., Longden, P.C. and Johnson, M.G., 1975. Insect pollination of sugar-beet (*Beta-Vulgaris*) seed crops. *Annals of Applied Biology,* **81**:127-134.

Garrett, M.H., Hooper, B.M., Cole, F.M. and Hooper, M.A., 1997, Airborne fungal spores in 80 homes in the Latrobe Valley, Australia: levels, seasonality and indoor – outdoor relationship, *Aerobiologia*, **13**:121-126.

Gloster, J., Champion, H.J., Mansley, L.M., Romero, P., Brough, T., Ramirez, A., 2005, The 2001 epidemic of foot-and-mouth disease in the United Kingdom: epidemiological and meteorological case studies. *Vet Rec.*, **156**:793-803.

Gottwald, T.R., Sun, X., Riley, T., Graham, J.H., Ferrandino, F. and Taylor, E.L., 2002, Geo-referenced spatiotemporal analysis of the urban citrus canker epidemic in Florida. *Phytopathology,* **92**:361-377.

Gregory, P.H., 1945, The dispersion of air-borne spores, *Trans. Br. mycol. Soc.*, **28**:26-72.

Gregory, P.H., 1951, Deposition of air-borne *Lycopodium* spores on cylinders, *Ann. appl. Biol.*, **38**:357-376.

Gregory, P.H., 1952, Spore content of the atmosphere near the ground, *Nature,* **170**:475.

Gregory, P.H., 1954, The construction and use of a portable volumetric spore trap, *Trans. Br. mycol. Soc.*, **37**:390-404.

Gregory, P.H., 1973, *The Microbiology of the Atmosphere*, 2nd edit, Leonard Hill, Aylesbury.

Gregory, P.H., 1976, *Outdoor Aerobiology*, Oxford University Press, London, pp 16.

Gregory, P.H., Festenstein, G.N., Lacey, M.E., Skinner, F.A., Pepys, J. and Jenkins, P.A., 1964, Farmers lung disease: the development of antigens in moulding hay, *J. gen. Microbial.*, **36**;429-439.

Gregory, P.H., Guthrie, E.J. and Bunce, M.E., 1959, Experiments on splash dispersal of fungus spores. *J. gen. Microbiol.*, **20**:328-354.

Gregory, P.H. and Henden, D.R., 1976, Terminal velocity of basidiospores ot he giant puffball (*Lycoperdon giganteum*), Trans. Br. mycol. Soc., **67**:399-407.

Gregory, P.H. and Hirst, J.M., 1952, Possible role of basidiospores as airborne allergens, *Nature*, **170**:414.

Gregory, P.H. and Hirst, J.M., 1957, The summer air-spora at Rothamsted in 1952, *J. Gen. Microbiol.*, **17**:135-152.

Gregory, P.H., Hirst, J.M. and Last, F.T., 1953, Concentrations of basidiospores of the dry rot fungus (*Merulius lacrymans*) in the air of buildings, *Acta allergologica*, **6**:168-174.

Gregory, P.H. and Lacey, M.E., 1963a, Mycological examination of dust from mouldy hay associated with farmer's lung, *J. gen. Microbiol.*, **30**:75-88.

Gregory, P.H. and Lacey, M.E., 1963b, Liberation of spores from mouldy hay, *Trans. Br. mycol Soc.*, **46**:73-80.

Gregory, P.H. and Lacey, M.E., 1964, The discovery of *Pithomyces chartarum* in Britain, *Trans. Brit. Mycol Soc.*, **47**:25-30.

Gregory, P.H., Lacey, M. E., Festintine, G.N. and Skinner, F.A., 1963, microbial and biochemical changes during the moulding of hay, *J. gen. Microbiol.*, **33**:147-174.

Gregory, P.H. and Read, D.R., 1949, The spatial distribution of insect-borne plant-virus diseases, *Ann. appl. Biol.*, **36**:475-482.

Gregory, P.H. and Sreeramulu, T., 1958, Air spora of an estuary, *Trans. Br. mycol. Soc.*, **41**:145-156.

Gregory, P.H. and Stedman, O.J., 1953, Deposition of air-borne *Lycopodium* spores on plane surfaces, *Ann. appl. Biol.*, **40**:651-674.

Guan, Y., Poon, L.L., Cheung, C.Y., Ellis, T.M., Lim, W., Lipatov, A.S., Chan, K.H., Sturm-Ramirez, K.M., Cheung, C.L., Leung, Y.H., Yuen, K.Y., Webster, R G. and Peiris, J.S., 2004, H5N1 influenza: a protean pandemic threat, *Proc Natl Acad Sci U S A.*, **101**:8156-8161.

Hamilton, E.M., 1959, Studies in the air spora, *Acta allerg.*, **13**:143-175.

Harries, M.G., Lacey, J., Tee, R.D. Cayley, G.R. and Newman Taylor, A.J., 1985, *Didymella exitialis* and late summer asthma, *The Lancet*, l:1065-1066.

Harris, W.F. 1955, *A Manual of the Spores of New Zealand Pteridophyta*, R.E. Owen, Wellington, pp186.

Hart, M. L., Wentworth, J. E. and Bailey, J. P., 1994, The Effects of Trap Height and Weather Variables on Recorded Pollen Concentration at Leicester. *Grana* **33**: 100-103.

Haskell, R.J. and Barss, H.P., 1939, Fred Campbell Meier, 1893-1938, *Phytopathology*, **29**:293-302.

Hawksworth, D.L., Kirk, P.M. Sutton, B.C. and Pegler, D.N., 1995, *Ainsworth and Bisby's Dictionary of the Fungi*, 8th ed., CAB International, Wallingfard, pp 616.

Hesse, W., 1884, Ueber quantitative Bestimmung der in der luft enthaltenen mikroorganismen, *Mitth. Kaiserl. Gesundheitsamte*, **2**:182-207.

Hesse, W., 1888, Bemerkungen zur quantitative bestimmung der mikroorganismen in der luft, *Z. Hyg. InfektKrankh.*, **4**:182-207

Hirst, J.M., 1952, An automatic volumetric spore trap. *Ann. appl. Boil.*, **39**:259-265.

Hirst, J.M., 1953, Changes in atmospheric spore content: diurnal periodicity and the effects of weather, *Trans. Br. mycol. Soc.*, **36**:375-393.

Hirst, J.M., 1990, Philip Herries Gregory: 24 July 1907 – February 1986, *Biog. Mems. Fell. R. Soc. Lond,* **35**:153-177.

Hirst, J.M., 1992, Biography of an aerobiologist: P.H. Gregory (1907–1986), *Aerobiologia,* **8**:209-218.

Hirst, J.M., 1994, Aerobiology at Rothamsted, *Grana,* **33**:66-70.

Hirst, J.M., 1995, Bioaerosols: introduction, retrospect and prospect, in: *Bioaerosols Handbook*, C.S. Cox and C M Wathes, Lewis publisher, Boco Raton, pp 1-10.

Hirst, J.M. and Stedman, O.J., 1963, Dry liberation of fungus spores by raindrops, *J. gen. Microbiol.,* **33** 335-344.

Hirst, J.M., Stedman, O.J. and Hogg, W.H., 1967a, Long-distance spore transport: methods of measurement, vertical spore profiles and the detection of immigrant spores, *J. gen. Microbiol.,* **48**:329-355.

Hirst, J.M. Stedman, O.J. and Hurst, G.W., 1967b, Long-distance spore transport: vertical sections of spore clouds over the sea, *J. gen. Microbiol.,* **48**:357-377.

Hirst, J.M. and Hurst, G.W., 1967, Long-distance spore transport, in *Airborne Microbes*, P.H. Gregory and J.L. Monteith, eds., *Symposium of the Society for general Microbiology*, **17**:307-344.

Hodgson, M.J. and Flannigan, B., 2001, Occupational respiratory disease: hypersensitivity pneumonitis, in: *Microorganisms in Home and Indoor Work Environments*, Flannigan, B., Samson, R.A. and Miller, J.D. eds., Taylor and Francis, London, pp 129-142.

Holliday, P. 1992 *A Dictionary of Plant Pathology*, Cambridge University Press, Cambridge, pp 369.

Hovmøller, M.S. Justesen, A.F. and Brown, J.K.M., 2002, Clonality and long-distance migration of *Puccinia striiformis* f.sp *tritici* in north-west Europe, *Plant Pathology* **51**:24-32.

Huber, L., Fitt, B.D.L. and McCartney, H.A., 1996, The incorporation of pathogen spores into rain-splash droplets: a modelling approach, *Plant Pathology,* **45**:506-517.

Hunter, R.G. and Lea, R.G., 1994, The airborne fungal populations of representative British homes, in: *Health Implications of fungi in indoor environments*, Samson, R.A., Flannigan, B., Flannigan, M.E., Verhoeff, A.P., Adan, C.C.G. and Moekstra, E.S., eds., pp 141-153.

Hyde, H.A. 1959. Volumetric counts of pollen grains at Cardiff, 1954-1957. *Journal of Allergy* **30**:219-234.

Hyde, H.A., 1972, Atmospheric pollen and spores in relation to allergy, *Clin. Allergy,* **2**:153-179.

Hyde, H.A. and Adams, K.F., 1958, *An Atlas of Airborne Pollen grains*, Macmillan, London, pp 112.

Hyde, H.A. and Adams, K.F., 1960. Counts of atmospheric pollen and spores at Cardiff as compared with those at Paddington, 1959. *International Archives of Allergy and Applied Immunology* **17**:250-251.

IMI, 1964 onwards, *IMI (CMI) Descriptions of Fungi and Bacteria*, CAB International, Wallingford.

Ingold, C. T., 1939, *Spore Discharge in Land Plants*, University Press, Oxford.

Ingold, C.T., 1971, *Fungal Spores: their Liberation and Dispersal,* Clarendon Press, Oxford.

Ingold, C. T., 1999, Active liberation of reproductive units in terrestrial fungi, *The Mycologist* **13:**113-116.

Jarosz, N., Loubet, B., Durand, B., McCartney, H.A., Foueillassar, X., and Huber, L., 2003, Airborne concentration and deposition rate of maize pollen, *Agriculture and Forest Meteorology*, **119**:37-51.

Kauppinen, E.I., Jappinen, A.V.K., Hillamo, R.E., Rantiolehtimaki, A.H. and Koivikko, A.S., 1989, A Static Particle-Size Selective Bioaerosol Sampler for the Ambient Atmosphere, *Journal of Aerosol Science*, **20** 829-836.

Kawashima, S., and Takahashi, Y., 1999, An improved simulation of mesoscale dispersion of airborne cedar pollen using a flowering-time map, *Grana* **38**:316-324.

Kennedy, R., Wakeham, A.J., Byrne, K.G., Meyer, U.M. and Dewey, F.M., 2000. A new method to monitor airborne inoculum of the fungal plant pathogens Mycosphaerella brassicicola and Botrylis cinerea. *Applied and Environmental Microbiology,* **66**:2996-3003.

Khandelwal, A., 1992, Airborne diatoms at Lucknow, India, *Ind. J. Aerobiol., Special Vol.,*:179-185.

Knight, T.A., 1799, *Experiments on the Fecundation of Vegetables*.

Lacey, J. 1971a, The microbiology of moist barley storage in unsealed silos, *Ann. appl. Biol.*, **69**:187-212.

Lacey, J. 1971b, Thermoactinomyces sacchari sp nov, Thermophilic actinomycete causing bagassosis, J Gen Microbiol, **66**:327-333.

Lacey, J., 1990, Isolation of thermophilic microorganisms, In: *Isolation of Biotechnological Organisms from Nature*, D.P.Labeda, ed., McGraw Publishing Co., New York, pp 141-181.

Lacey, J., 1996, Spore dispersal – its role in ecology and disease: the British contribution to fungal aerobiology, *Mycol. Res.*, **100**:641-660.

Lacey, J., 1997, Actinomycetes in composts, *Ann. Agric. Environ. Med.*, **4**:113-121.

Lacey, J. and Crook, B., 1988, Fungal and actinomycete spores as pollutants of the workplace and as occupational allergens, *Ann. occup. Hyg.*, **32**:515-533.

Lacey, J. and Dutkiewicz, J., 1994, Bioaerosols and occupational lung disease, *J. Aerosol. Sci.*, **25**:1371-1404.

Lacey, J. and Lacey, M.E., 1964, Spore concentrations in the air of farm buildings, *Trans Br. mycol. Soc.*, **47**:547-552.

Lacey, J. Lacey, M.E. and Fitt, B.D.H., 1997, Philip Herries Gregory 1907-1986: pioneer aerobiologist, versatile mycologist, *Annu. Rev. Phytopathol.*, **35**:1-14.

Lacey, J., Pepys, J. and Cross, T., 1972, Actinomycetes and fungus spores in air as respiratory allergens, in: *Safety in microbiology*, D.A. Shapton and R.G. Board, eds., Academic Press, London, pp 151-184.

Lacey, J. and Venette, J., 1995, Outdoor air sampling techniques, in: *Bioaerosols Handbook,* C.S. Cox and C.M. Wathers, eds., Lewis Publishers, Boca Raton, pp 407-472.

Lacey, J., Williamson, P.A.M. and Crook, B., 1992, Microbial emissions from composts made for mushroom production and from domestic waste, in: *Composting and Compost Quality Assurance Criteria*, D.V. Jackson, J.M. Merillot and P. L'Hermite, eds. Luxembourg, office for Official Publications of the European Community. EUR14254, pp 117-130.

Lacey, M.E., 1962, The summer air-spora of two contrasting adjacent rural sites, J. Gen. Mycrobiol., **29**:485-501.

Lacey, M.E., 1997, Painting as an aid in identifying fungal spores. In: *Aerobiology: Proceedings 5th Aerobiological Conference, Bangalore.* in: S.N. Agashe, ed., Oxford & IBH Publishing CO. PVT. Ltd., New Delhi, pp 7-12.

Lacey, M.E. and McCartney, H.A., 1994, Measurement of airborne concentrations of spores bracken (*Pteridium aquiiinum*), *Grana*, **33**:91-93.

Lacey, M.E, Rawlinson, C.J. and McCartney, H.A., 1987, First record of the natural occurrence in England of the teleomorph of *Pyrenopeziza brassicae* on oilseed rape, *Trans. Br., mycol. Soc.*, **89**:135-140.

Last, F.T., 1955, Spore content of air within and above mildew-infected cereal crops, *Trans. Br. mycol. Soc.*, **38**:453-464.

Legg, B.J., 1983, Movement of plant pathogens in the crop canopy, *Phil. Trans. R. Soc. Lond.*, **B 302**:559-574.

Legg, B.J. and Bainbridge, A. 1978, Air movement within a crop: spore dispersion and deposition, In: *Plant Disease Epidemiology*, Scott, P.R. and Bainbridge, A., eds., Blackwell Scientific Publications, Oxford.

Limpert, E., Godet, F., and Muller, K., 1999, Dispersal of cereal mildews across Europe, *Agricultural and Forest Meteorology.* **97:**293-308.

Linskens, H.F., Bargagli, R., Cresti, M. and Focardi, S., 1993. Entrapment of long-distance transported pollen grains by various moss species in coastal Victoria Land, Antarctica, *Polar biology,* **13**:81-87.

Macdonald, O.C. and McCartney, H.A., 1987, Calculation of splash droplet trajectories, *Agriculture and Forest Meteorology*, **39**:95-110.

Mackay, T.W., Wathen, C.G., Sudlow, M.F., Elton, R.A. and Caulton, E., 1992, Factors affecting asthma mortality in Scotland, *Scottish Medical Journal,* **37**: 5-7

Maddox, R.L., (1870), On an apparatus for collecting atmospheric particles, *Monthly Microsc. J.*, **3**:286-290.

Marshall, W.A., 1996. Biological particles over Antarctica. *Nature,* **383**: 680.

May, K.R., 1945, The cascade impactor: an instrument for sampling coarse aerosols, *J. Sci. Instrum.*, **22**:187-195.

McCartney, H.A., 1987, Deposition of *Erysiphe graminis* conidia on a barley crop II. consequences for spore dispersal, *J. Phytopathology*, **118**:258-264..

McCartney, H.A., and Aylor, D.E., 1987, Relative contributions of sedimentation and impaction to deposition of particles in a crop canopy, *Agricultural and Forestry Meteorology*, **40**:343-358.

McCartney, H.A. and Bainbridge, A., 1987, Deposition of *Erysiphe graminis* conidia on a barley crop I. sedimentation and impaction, *J. Phytopathology*, **118**:243-257.

McCartney, H.A. Bainbridge, A. and Stedman, O.J., 1985, Spore deposition velocities over a barley crop, *Phytopath. Z.,* **114**:224-233.

McCartney, H.A. and Fitt, B.D.L., 1985, Construction of dispersal models, in, *Advances in Plant pathology Vol. 3: Mathematical Modelling of Crop Disease*, C. A. Gilligan, ed., Academic Press, London, pp 107-143.

McCartney, H.A., Fitt, B.D.L. and Schmechel, D., 1997, Sampling bioaerosols in plant pathology, *Journal of Aerosol Science*, **28**:349-364.

McCartney, H.A. and Lacey, M.E., 1990, The production and release of ascospores of *Pyrenopeziza brassicae* on oilseed rape, *Plant Pathology*, **39**:17-32.

McCartney, H.A. and Lacey, M.E., 1991a, The relationship between the release of ascospores of *Sclerotinia sclerotiorum*, infection and disease in sunflower plots in the United Kingdom, *Grana*, **30**:486-492.

McCartney, H.A. and Lacey, M.E., 1991b, Wind dispersal of pollen from crops of oilseed rape (*Brassica napus* L), *J. Aerosol Sci.*, **22**:467-477.

McCartney, H.A. and Lacey, M.E., 1999, Timing and infection of sunflowers by *Slerotinia sclerotiorum*, in: *Protection and Production of Combinable Break Crops, Aspects of Applied Biology*, **56**:151-156.

McCartney, H.A., Lacey, M.E. and Rawlinson, C.J., 1986, Dispersal of *Pyrenopezziza brassicae* spores from oil-seed rape crop, *J. Agric. Sci.*, **107**:299-305.

McCartney, H.A., Schmechel, D. and Lacey, M.E., 1993, Aerodynamic diameter of *Alternaria* species, *Plant Pathology*, **42**:280-286.

Meier, F.C., 1935, Micro organisms in the atmosphere of the arctic regions, *Phytopathology*, **25**: 27.

Mercer, T.T., 1973, *Aerosol Technology in Hazard Evaluation*, Academic Press, New York.

Mims, S.A. and Mims, F.M., 2004. Fungal spores are transported long distances in smoke from biomass fires. *Atmospheric Environment*, **38**:651-655.

Miquel, P., 1883, *Les organismes vivants de l'atmosphère*, Gauthier-Villars, Paris, pp 310

Moore, P.D., Webb, J.A. and Collinson, M.E., 1991, *Pollen Analysis*. 2nd ed, Blackwell Scientific Publications, Oxford, pp 216.

Morris, C.W., Boddy, L., and Allman, R., 1992, Identification of Basidiomycete Spores by Neural Network Analysis of Flow-Cytometry Data, *Mycological Research*, **96**:697-701.

Morrow-Brown, M., 1994, The relationship between aerobiology and allergology, in: *Recent Trends in Aerobiology and Immunology: a Collection of plenary Lectures and Contributory Articles.*, Agashe, S., ed., Oxford & IBH Publishing CO. PVT., New Delhi, pp 1-30.

Moulton, S. (ed.), 1942, *Aerobiology. Amer. Assoc. Adv. Sci.*, Pub. No. 17, Washington, pp 289.

Mullins, J., 2001, Microorganisms in outdoor air in: *Microorganisms in Home and Indoor Work Environments.* B. Flannigan, R.A. Samson and J.D. Miller, eds., Taylor & Francis, London, pp 3-16.

Mullins, J. and Emberlin, J., 1997, Sampling pollens, *J. Aerosol Sci.*, **28**:365-370.

Newson, R., Strachan, D., Corden, J. and Millington, W., 2000, Fungal and other spores as predictors of admissions for asthma in the Trent region, *Occupational and Environmental Medicine*, **57**: 786-792.

Nikkels, A.H., Terstegge, P., and Spieksma, F.Th.M., 1996, Prevalence of fungi in carpeted floor environments: analyses of dust samples from living rooms, bedrooms, offices and school classrooms, *Aerobiologia*, **12**: 113-120.

Nilsson, S., ed. 1983 *Atlas of Airborne Fungal Spores in Europe.* Springer-Verlag, Berlin, pp 139.

Nilsson, S. Praglowski, J. and Nilsson, L., 1977, *Atlas of Airborne Pollen Grains and Spores in Northern Europe,* Natur och Kultur, Stockholm, pp 3-16.

Pasteur, L., 1861, Mémoire sur les corpuscles organisés qui existent dans l'atmosphère. Examen de la doctrine des generations spontanées, *Ann. Sci. Nat. (Zool.),* 4 sér., **16**:5-98.

Pehkonen, E., Rantiolehtimaki, A., 1994. Variations in Airborne Pollen Antigenic Particles Caused by Meteorological Factors. *Allergy* **49**:472-477.

Pepys, J., Jenkins, P.A., Festentein, G.N., Gregory, P.H., Lacey, M.E., and Skinner, F.A., 1963, Farmer's lung: thermophilic as a source of 'farmer's lung hay' antigen, *The Lancet* ii, 607-611.

Perkins, W.A. 1957, The rotorod sampler, 2nd *Semiannual Rept. Aerosol Lab.,* Dept. Chem. and Chem. Engng., Stanford University. CML., 186,

Pielaat, A., van den Bosch, F., Fitt, B.D.L., and Jeger, J., 2001, Simulation of vertical spread of plant diseases in crop canopy by stem extension and splash dispersal, *Ecological Modelling*, **151**:199-216.

Postgate, J., 1986, *Microbes and Man*, Penguin books, Harmondsworth, pp 239.

Prodi, V., Melandri, C., Tarroni, G., Dezaiacomo, T. and Formignani, M. 1979, Inertial Spectrometer for Aerosol-Particles. *Journal of Aerosol Science* **10**:411-419.

Rantio-Lehtimäki, A., 1991, Aerobiology in Finland. 1. Information Service, *Methodology and Allergolological Applications*, Ph.D. Thesis, University of Turku.

Reponen, T., 1994, Viable fungal spores as indoor aerosols, Ph.D. Thesis, University of Kuopio.

Samson, R.A., Houtbraken, J., Summerbell, R.C., Flannigan, B. and Miller, J.D., 2001, Common and important species of fungi and actinomycetetes in indoor environments, in: *Microorganisms in home and indoor work environments*, Flannigan, B., Samson, R.A. and Miller, J.D. eds. pp 287-473.

Sache, I., 2000, Short-distance dispersal of wheat rust spores by wind and rain, *Agronomie*, **20**:757-767.

Sánches Mesa, J.A., Smith, M., Emberlin, J., Allitt, U., Caulton, E. and Galan, C., 2003, Characteristics of grass pollen season in areas in southern Spain and the United Kingdom, *Aerobiologia*, **19**:243-250.

Schlichting, H. E., 1971, Preliminary Study of Algae and Protozoa in Seafoam. *Botanica Marina*, **14**:24-29.

Schlichting, H. E., 1974, Ejection of Microalgae into Air Via Bursting Bubbles. *Journal of Allergy and Clinical Immunology* **53**:185-188.

Schmechel, D., McCartney, H. A., Magan, N., 1996, A novel approach for immunomonitoring air-borne fungal pathogens. In *Diagnostics in Crop Production* (Edited by Marshall, G.) British Crop Protection Council, Farnham.

Schmidt, W., 1925, Der Massenaustausch in frier Luft und verwandte Erscheinungen, *Probl. Kosm. Phys.*, **7**:1-118.

Schneider, J.M., Roos, J., Lubin, Y. and Henschel, J.R., 2001, Dispersal of *Stegodyphus dumicola* (Araneae, Eresidae): They do balloon after all! *Journal of Arachnology* **29**:114-116.

Scott, R. K., 1970, Effects Weather on Concentration of Pollen within Sugar-Beet Seed Crops. *Annals of Applied Biology*, **66**:119-127.

Shaw, R.H. and McCartney, H. A., 1985, Gust penetration into plant canopies, *Atmospheric Environment*, **19**:827-830.

Simán, S.E., Povey, A.C. and Sheffield, E., 1999, Human health risks from fern spores? a review, *Fern Gazette*, **15**:275-287.

Simons, P., 1996, *Weird Weather*, Warner books, London, pp 309.

Skaug, M.A., Eduard, W. and Stormer, F.C., 2001, Ochratoxin A in airborne dust and fungal conidia. *Mycopathologia*, **151**:93-98.

Spieksma, F.Th.M. Corden, J.M., Detandt M., Millington,W.M., Nikkels, H. Nolard, N., Shoenmakers C.H.H., Wachter, R., de Weger, L.A., Willems, R and Emberlin, J, 2003, Quantitative trends in annual totals of five common airborne pollen types (*Betula, Quercus,* Poaceae, *Urtica* and *Artemesia*) at five pollen-monitoring stations in western Europe, *Aerobiologia*, **19**: 171-184.

Spiewak, R., ed. 1995, *Pollens and pollinosis current problems*, Institute of Agricultural Medicine, Lublin.

Sreeramulu, T. and Ramalingam, A., 1966, A two year study of the air spora of a pady field near Visakhapatnam, *Indian J. agric. Sci.*, **36**:111-132.

Stedman, O.J., 1980a, Splash droplet and spore dispersal studies in field beans (Vicia faba L), *Agricultural Meteorology*, **21**:111-127.

Stedman, O.J., 1980b, Splash dispersal studies in wheat using a fluorescent tracer, *Agricultural Meteorology*, **21**:195-203.

Stern, M.A., Allitt, U., Corden, J. and Millington, M., 1999, The investigation of fungal spores in intramural air using a Burkard continuous recording sampler, *Indoor Built Environment* **8**:40-48.

Stepanov, K.M., 1935, Dissemination of infective diseases of plants by air currents,(In Russion, English title) *Bull. Pl. Prot. Leningr.,* Ser. 2, *Phytopathology,* no. **8**:1-68.

Sutton, O.G., 1932, A theory of eddy diffusion in the atmosphere, *Pro. R. Soc* A., **135**:143-165.

Swan, J.R.M., Gilbert, E.J., Crook, B., 2002, Microbial emissions from composting sites, In: *Issues in Environmental Science; 18, Environmental and Health impact of solid waste management activities,* Hester, R.E., and Harrison, R.M., eds., RSC Press, Cambridge, pp 73-101.

Tervet, I. W., 1950, A technique for the collection of dry spores from infected plants. (Abstr.) *Phytopathology*, **40**:874.

Tervet, I. W., Rawson A. J., Cherry E. and Saxon R. B., 1951, A method for the collection of microscopic particles *Phytopathology*, **41**:282-285.

Thorpe, A., Gould, J.R., Brown, R.C. and Crook, B., 1997, Investigation of the performance of agricultural vehicle cab filtration systems against grain dust, *J. Agric. Eng. Research,* **66**:135-149.

Tilak, S.T., 1989, *Airborne Pollen and Fungal spores*, Vaijayanti Prakashan, Aurangabad, pp 316

Tilak, S.T. and Kulkarni, R. L. 1970, A New Air Sampler. *Experientia*, **26**:443-447.

van Woerden, H.C., Mason, B.W., Nehaul, L.K., Smith, R., Salmon, R.L., Healy, B., Valappil, M., Westmoreland, D., de Martin, S., Evans, M.R., Lloyd, G., Hamilton-Kirkwood, M., and Williams, N.S.,2004, Q fever outbreak in industrial setting. *Emerg. Infect Dis.*, **10**:1282-1289.

Verhoeff, A.P., 1994, Home dampness fungi and house dust mites and respiratory symptoms in children, Ph.D. Thesis, University, Rotterdam.

Wakeham, A., Kennedy, R. and McCartney, A., 2004, The collection and retention of a range of common airborne spore types trapped directly into microtiter wells for enzyme-linked immunosorbant analysis, *Aerosol Science*, **35**:835-850.

Walklate, P.J.,1989, Vertical dispersal of plant pathogens by splashing. Part I: relationship between rainfall and upward rain splash, *Plant Pathology*, **38**:56-63.

Walklate, P.J.,McCartney, H.A. and Fitt, B.D.L., 1989, Vertical dispersal of plant pathogens by splashing. Part II: experimental study of the relationship between raindrop size and the maximum splash height, *Plant Pathology*, **38**:64-70.

Wang, B., Zhang, A., Sun, J.L., Liu, H., Hu, J. Xu, L.X., 2005, Study of SARS transmission via liquid droplets in air. *J Biomech Eng.*, **127**:32-38.

West, J.S., Bravo, C., Oberti, R., Lemaire, D., Moshou, D. and McCartney, H.A., 2003, The potential of optical canopy measurement for targeted control of field crop diseases, *Annual Review of Phytopathology* **41**:593-614.

West, J.S., Fitt, B.D.L., Leech, P.K., Biddulph, J.E. and Huang, Y.J., 2002b, Effects of timing of *Leptoshaeria maculans* ascospore release and fungicide regime on phoma leaf spot and phoma stem canker development on winter oilseed rape (*Brassica napus*) in southern England, *Plant Pathology*, **51**:454-463.

West, J.S., Jedryczka, M., Leech, P.K., Dakowska, S., Huang, Y.J., Steed, J.M. and Fitt, B.D.L., 2002a, Biology of *Leptosphaeria maculans* ascospore release in England and Poland, *IOBC/wprs Bulletin* Vol: **25**:21-29. Presented at the IOBC conference, Soest, Germany, April 2001.

West, J.S. and McCartney, H.A., 2003, Predicting latent infection around stripe rust foci for spatially variable fungicide application, *8th International Congress of Plant Pathology, 2-7 February, 2003, Christchurch, New Zealand.* Abstract 8.51, **2**:113.

Weyman, G. S., Sunderland, K.D. and Jepson, P. C., 2002, A review of the evolution and mechanisms of ballooning by spiders inhabiting arable farmland. *Ethology Ecology & Evolution*, **14**:307-326.

Williams, R.H., Ward, E. and McCartney, H.A., 2001, Methods for integrated air sampling and DNA analysis for detection of airborne fungal spores, *Applied and Environmental Microbiology*, **67**:2453-2459.

Willis, J.C., 1960, *A Dictionary of the Flowering Plants and Ferns*, The University Press, Cambridge.

Yu, I.T., Li, Y., Wong, T.W., Tam, W., Chan, A.T., Lee, J.H., Leung, D.Y. and Ho, T., 2004, Evidence of airborne transmission of the severe acute respiratory syndrome virus. *N Engl J Med.*, **350**:1731-1739.

Index

abiotic partles 93, 125
Absidia corymbifera 109
A. ramose 109
Acacia 77
A. auriculiformis 99
Acalypha 101
Acanthaceae 101
Acer pseudoplatanus 101
Aceraceae 101
Acremoniella atra 111
actinomycetes 2, 3, 27, 29, 80, 94, 125, 135
adhesive 39, 42, 43, 50, 52, 55, 58, 63
Aecidiospore 105
Aecidium mori 105
aerobiology XIII, 1, 2, 4, 9, 12, 13, 14, 15, 22, 25, 28, 80, 125
aerobiology - pathway 9, 12, 13, 15-34
aerodynamic 8, 19, 20, 21, 35
– diameter 20, 21, 35, 125
– effects 8
aerosol 13, 15, 18, 29, 30, 31, 34, 35, 125
–, bioaerosol 14, 35, 46, 125
agar 35, 39
Agaricus 16, 107
A. bispora 107
agriculture 6
air
– filters 21, 29
– intake 36, 37, 38, 43, 44, 45, 51, 54, 82
– mass movement 19, 20
– movement 16, 18, 19, 35, 40, 54
– resistance 20
–, volume of 18, 42, 56, 73, 87, 89
–, volume of air sampled 43, 87, 89, 90
– conditioning system 30
air sampler
– –, aeroconiscope 6, 7
– – Andersen sampler 21, 24, 27, 28, 29, 31, 35, 38, 39
– –, Biotrace intelligent cyclone air sampler 44
– –, Burkard portable air sampler 39
– –, Burkard trap 21, 27, 30, 31, 35, 36, 40, 43, 49-58, 68, 74, 75, 77, 79, 81, 82, 83, 118, 120, 125
– –, cascade impactor 9, 10, 24, 27, 28, 35, 37, 38, 79, 126
– –, cyclone 12, 13, 35, 36, 44, 45, 47
– –, cyclone respirable dust sampler 47
– –, Hirst spore trap 10, 11, 36, 43, 93, 126
– –, impactor 9, 10, 12, 21, 24, 27, 28, 30, 33, 35, 36, 37, 38, 45, 46, 79, 127
– –, impactor, 7-stage 79
– –, impinger 45, 46, 127
– –, jet spore sampler 12
– –, Lanzoni 11, 44
– –, liquid impinger 31, 45, 46
– –, MAQS II microbiological 35
– –, MAS 100, 35
– –, MTIST 36
– –, passive
– – –, Petri dish 9, 36
– – –, sticky rod/cylinder 9, 10, 31, 37
– – –, sticky slide 9, 31, 35
– – –, sticky strip 37
air sampler
– –, rotating arm 11, 12, 13, 21, 41
– –, Rotorod 11
– –, size-selective bioaerosol sampler (SSBAS) 46
– –, SKC BioSampler 36
– –, Tilak 44
– –, VersaTrap 36
– –, whirling arm trap 11, 21, 31, 35, 39, 40, 41, 43, 73, 79, 89, 119, 120, 123
air
– sampling 9, 11, 12, 15, 21, 31, 35, 45, 47
– spora XIII, 1, 2, 3, 5, 8, 10, 11, 17, 24, 28, 35, 43, 49, 73, 83, 91, 125
–, still 4, 16, 20, 33, 45
–, upper 9
airborne

– bacteria 8, 24, 29
– basidiospores
– biological particles XIII, 1, 11, 35, 36
– dispersal 4
– germs 8
– microbes 29
– particles 11, 12, 13, 21, 22, 24, 35, 38, 43, 44, 50, 73, 84, 91, 94
– pollen 17, 18, 93
– spores 4, 9, 16, 22, 24, 29, 30, 31, 93
aircraft 32
airstream 36, 37, 46, 47
Albugo 109
alder 99
algae 2, 3, 18, 92, 112, 113, 125
Alice Springs 77
allergen XIII, 15, 46, 47, 93, 125
allergenic 11, 14, 25, 26, 93
allergenicity 35
allergic reactions 1, 27, 29, 93
allergy 8, 14, 17, 21, 22, 24, 25, 49, 94
Alnus glutinosa 99
Alopecurus myosuroides 92, 97
Alps 4
Alternaria 10, 17, 27, 32, 77, 89, 113
A. alternate 113
A. brassicae 113
A. brassicicola 113
A. infectoria 113
alveolitis 25, 27, 28, 111
Amanita muscaria 107
A. rubescens 107
Amaranthaceae 99
Amaranthus viridis 99
Ambrosia deltoidea 101
amethist deceiver 107
amoeboid cyst 115
Anacardiaceae 101
anamorph 125
anemophilous plants 3, 17, 125
angiosperm 3, 7, 125
animal XIII, 1, 3, 6, 15, 18, 28, 30, 93, 111
– feed 28
– material 93
annulus 18
Antarctica XIII, 33, 94, 113
anthers 17
Anthiscus sylvestris 101
anthrax 6
apothecia 17, 22, 23, 31, 125
apple scab 17, 105, 111

Arctic 32
Areca catechu 97
Armillaria mellea 107
armoured cable, 49, 50
Artemisia vulgaris 101
Arthrinium arundinis 113
ascomycete fungi 31, 92, 125
ascospore 10, 17, 21, 22, 27, 31, 75, 79, 92, 105
– release 22, 23
ascus 17, 31, 92, 125
Asia 32
aspergilloma 29
aspergillosis 29, 111
Aspergillus 79, 92
A. fumigatus 27, 29, 111
A. glaucus 111
A. niger 111
Asplenium nidus 103
Asterosporium 111
asthma 11, 14, 24, 25, 26, 27, 28, 105, 111, 120
atmosphere XIII, 2, 3, 4, 5, 6, 8, 94, 125
Atriplex canescens 99
Australia 22, 77, 97, 103, 105, 115
avian 'flu 30
Azadirachta indica 101

Bacillus 33. 115
B. subtillus 115
backtracking 31
back-trajectory analysis 32, 33
bacteria ,bacterial XIII, 1, 2, 3, 8, 13, 15, 22, 24, 26, 27, 29, 30, 32, 33, 34, 79, 80, 92, 115, 125
Badhamia utricularis 109
bagassosis 28, 115
ballistic drop 34
ballistospore, 16, 111
balloon 18, 20
banana plantation 75
Barbula fallax 113
bare-toothed Russula 107
barley 12, 17, 27, 31, 36, 111
barley and rye leaf blotch 111
barometric pressure 2
Bartramina patens 113
basic fuchsin 57, 118
basidiomycetes 92, 125
basidiospore 11, 75, 92, 107, 125
basidium 92, 125
bats 2
battery 11, 39, 42, 43, 44, 50, 54
beads 12

Beagle, The 5, 6
beans 105
bedding 24, 30
beech 99
beefsteak fungus 107
Beltrania 111
Benninghoff 12
betel nut 97
Betula 18, 99
B. verrucosa 99
Betulaceae 99
bioaerosol 14, 35, 46, 125
biochemical
– analysis 46
– changes 27
biomass fires 32
birch 105
– allergen 46
birds 2
Bispora monilioides 111
black grass 97
Black Sea 33
Blackley 8
Blastomyces dermatitidis 111
blastomycosis 111
Blumaria 77
B. graminis 17, 111
Bolbitius vitellinus 107
Boletus chrysenteron 107
B. elegans 107
B. scaber 107
Borassus flavellifer 97
Bordetella pertussis 30
Botryodiplodia 111
B. acerina 111
Botrytis 77, 111
bottle brush 101
boundary layer 125
– – laminar 2, 3, 4, 16, 21, 127
– – planetary 2
– – turbulent 2, 3, 16, 49
bovine mycotic abortion 109
Bovista plumbea 107
bracken 30, 77, 81, 103
brassica
– canker 17, 22, 79, 105, 111
– leaf spots 105, 111, 113
Brassica napus 20, 77, 99
Brisbane 77
British Aerobiology Federation (BAF) XIII, 13, 50, 52, 55, 60, 61, 63, 64, 70, 72, 82, 83, 85, 86, 120

Brompton Hospital 27
bronchopulmonary aspergillosis 29
brown spot of rice 113
Bryum algens 113
bubble burst 15, 18, 33
buck's-horn plantain 101
building 2, 3, 8, 9, 24, 28, 30, 34
Buller 5, 9
bur reed 97
butt rot 107

cable 43, 49, 50
Cadham 8
Caeoma euphorbiae-geniculate 105
Calcutta 6, 7, 19
calibration 10, 38, 69, 70, 71, 80, 91, 126
Callistemon citrinus 101
Calluna vulgaris 101
Calvatia gigantia 107
camera 65, 94
Cape Verde Islands 5
Caprifoliaceae 101
capsule 18, 92
carbon dioxide 2
carbon shards 115
carcinogenic 30
Cardiff 11, 17
Carex nigra 97
Carica papaya 101
Caricaceae 101
Carpinus betulus 99
Casella Ltd. 11
Castanea sativa 99
Casuarina equisetifolia 94, 97
C. cunninghamiana 94, 97
Casuarinaceae 97
cat fur 115
cattle 3, 109
Cecropia 99
cedar of Lebanon 103
Cedrus libani 103
Ceratodon purpureus 113
Ceratosporiella 113
Cercospora 111
C. arachidicola 111
Cercosporidium personatum 111
cereals 105, 109, 111
Chaetomium globosum 105
Chamaenerion augustifolium 20
Chamberlain, 19, 21, 36

chemical analysis 39, 50
Chenopodiaceae 99
Chenopodiun album 99
chimney 18, 19
chocolate spot of beans 105
cholera 6
Chondrostereum purpureum 107
Chrysosporium 111
citrus canker 22, 34
city dwellers 25, 26
Cladonia 92, 113
Cladosporium 10, 16, 17, 21, 27, 32, 75, 77, 79, 113
C. cladosporioides 113
C. herbarum 113
classification XV
Claviceps purpurea 17, 105
clay 4, 33
Clematis vitalba 20
climate change 15
climatic factors 23
clock 10, 51, 53, 54
clubmoss 103
clumping of particles 8, 9, 31
clustered tough-shank 107
Coccidioides immitis 111
coccidiomycosis 111
coconut palm 97
Cocos nucifera 97
collecting
– liquid 44
– medium 36
– surface 36, 40, 89
collection chamber 45, 46
Collybia confluens 107
C. maculate 107
colonies 8, 13, 31, 38
colonisation of new habitats 15, 21
combine harvester 28, 29
combustion product 115
common cold 30
common ink cap 107
common oak 99
common yellow Russula 107
community health 25
Compositae 101
compost 24, 29
compound microscope 59, 61, 62
concentration XIII, 3, 5, 13, 16, 18, 19, 20, 22, 28, 29, 31, 32, 36, 37, 62, 73, 80, 81, 83, 84, 86, 87, 88, 89, 126

–, bacteria 29
–, bioaerosols 13
–, particles 3, 5, 18, 73, 80, 83, 84
–, pollen 18, 19, 20, 32, 86-89
–, spores 16, 19, 20, 22, 28, 32, 36, 37, 86-89
–, standard measurement 10, 126
condenser 59, 62, 65, 66, 67
Conidiobolus obscure 109
conidium, conidia 31, 33, 75, 79, 92, 109, 111, 113, 126
conifer 17, 22, 103, 107, 126
– polypore root rot 107
coniferous pollen 103
contamination 2, 44, 91
continents 12
convection 19, 20, 32
cooling systems 30
Coprinus atramentarius 107
C. micaceus 107
coral spot 105, 113
cork factory 79
cork oak 99
correction factor 73, 84, 86, 87, 88, 89, 121, 122, 123
Cortinarius elatior 107
Corylaceae 99
Corylus avellana 99
Corynespora 111
Costa Rica XIII, 75, 105, 111, 223
cotton 5, 115
cotton blue 57, 80, 118
counting
– convention 80, 81, 86
– season 88
– sheets, 84, 86, 120, 121, 122, 123
countries 11, 12, 30, 49
counts 11, 18, 24, 25, 27, 49, 56, 80, 82, 83, 84, 86, 87, 88, 126
cover slip 43, 56, 57, 63, 65, 66, 67, 69, 70, 85
covered smut of sorghum 109
cow parsley 101
Coxiella burnetii 30
Crassulaceae 99
Craterellus cornucopioides 107
Crepidotus mollis 107
crop - canopy 19, 33, 36
crop 8, 11, 12, 14, 15, 17, 18, 19, 20, 22, 23, 29, 31, 33, 34, 36, 41, 50, 75, 77
cross pollination 19
Crucibulum vulgare 16
Cruciferae 99

Cryptostroma corticale 113
cultivated mushroom 107
culture 8, 35, 46, 94, 126
culturing 13, 46
Cunningham 6, 7
Cupressaceae 103
curled dock 99
Curvularia 32, 113
C. lunata 113
cyclone (*see*: air samplers) 126
cyclone (weather) 33, 126
Cyclotella 113
Cylindrosporium concentricum 105, 111
Cyperaceae 97
cyst 93, 115

daily
– counts 49, 83, 84, 126
– periodicity 17, 126
Daldinia concentrica 105
dandelion 20, 101
dander 26
Darwin 5, 6
date palm 97
day 2, 32
debris XIII, 1, 8, 15, 17, 22, 31, 79
decomposition 2
dehiscence of sporangia 18
Deightoniella torulosa 16
Delonix regia 99
density 2, 19, 20, 21, 32, 47, 91
deposition 2, 3, 5, 9, 10, 12, 13, 18, 19, 20, 21, 36, 37, 39, 49, 82, 83, 85, 126
desiccation 22
detection 12, 13, 15
dew 12, 17
diatom 2, 5, 18, 33, 92, 93, 113, 126
dicotyledons 94, 96, 97, 98, 99, 101, 126
Dicranopteris linearis 103
Dictydium 16
Didymella 105
D. exitialis 27
Didymosphaeria donacina 105
diesel particles 115
diffusion 9, 126
discharge 16, 32
discomycete 16
disease 1, 3, 4, 5, 6, 8, 9, 15, 21, 22, 23, 24, 25, 27, 28, 29, 30, 31, 33, 34, 94
–, bacterial 8, 24, 30, 34
– control 22

– foci 23, 24
– forecasting 9, 15
– gradient 23, 24
– symptoms 22, 23, 28
diseases
–, animal 1, 3, 6, 21, 24, 30, 94, 109, 111
–, human 1, 6, 8, 21, 24, 27, 28, 29, 30, 94, 111
–, plant 1, 6, 8, 9, 15, 21, 22, 31, 33, 94, 105, 107, 111, 113
dispersal (dispersion) XIII, 1, 4, 5, 8, 9, 11, 12, 13, 14, 15, 16, 18, 19, 20, 21, 22, 23, 31, 32, 33, 34, 50, 83, 126
– gradient. 20, 23
–, long distance 5, 12, 19, 30, 31, 32, 33
– of pollen 1, 9, 11, 13, 19, 20, 31
– of spores 4, 8, 9, 16, 18, 19, 31, 33
dissecting needle 51, 53
distribution maps 93
diurnal periodicity 10, 17, 84
diversity 4, 10, 12, 15, 24, 43, 73, 91, 93
DNA 12, 35, 36, 47, 127
dogs 26
domestic waste 29
Dominican Republic 32
Dorset 27
downwind 20, 32, 50
downy mildew 109
drag forces 20
drawing tube 65, 94
Drechslera 77, 113
D. oryzae 113
drop 16, 21, 33, 34, 57, 67, 69
droplet - splash 34
drum 27, 28, 43, 50, 51, 52, 53, 54, 55, 85
dry conditions 18, 31
dry rot 11, 107
drying, 17, 18, 92
dust 2, 4, 5, 6, 8, 24, 26, 27, 28, 29, 47, 64, 65 69, 83, 93
– clouds 38
– extraterrestrial 4

earth fan 107
earth's - surface 2, 4
earth's atmosphere 2, 4
ecology 1, 14, 73
eczema 25, 31
eddy 2, 4, 18, 24, 126
– diffusion 9
Edmunds 12
Ehrenberg 5, 6

Elaeis guineensis 97
elater 18, 103
elder 101
ELISA 13, 126
elm 99
England 4, 4, 19, 31, 32
Engler's classification 94
Enteridium lycoperdon 109
Entoloma rhodopolium 107
entomophilous plants 17, 126
Entomophthora 16
E. planchoniana 109
environmental
– condition 3
– factors 3, 12
enzyme-linked immunosorbent assay (ELISA) 13
Epicoccum 77, 113
epidemiology 9, 15, 22, 31
Eppendorf tubes 12, 45
equations 40, 89
Equisetum 18, 103
ergot 17, 105, 113
Ericaceae 101
Erynia neoaphidis 109
Erysiphe 17, 111
estuary, 11
Eucalyptus 101
Euphorbia hirta 101
Euphorbiaceae 101
Europe, 4, 12, 22, 93
exine 92
Exosporium 113
extraction 13
eyepiece 59, 60, 61, 62, 63, 64, 66, 67, 69, 70, 71, 72, 80, 88, 89, 91, 121, 122, 123

facial eczema of sheep 111
faecal
– material 2
– pellets 93
Fagaceae 99
Fagus sylvatica 99
fall speed 20, 126
fan 28, 107
farm buildings 28
farmer's lung 24, 27, 28
fat-hen 99
fauna 1
feather 115
fermentation 2
fern 2, 4, 18, 30, 92, 93, 103, 126

fern - spore 103
fever 4, 6, 8, 14, 15, 24, 30
field 11, 23, 31, 37, 45, 50, 59, 62, 65, 66, 67, 71, 73, 79, 80, 88, 90, 97
field of vision 59, 80,
field woodrush 97
Fienberg
filiform 33
filter 5, 13, 14, 19, 29, 31, 34, 39, 45, 46, 47, 49, 62, 126
filtration 46, 47, 126
fire 4, 27, 32, 115
fish 2
Fistulina hepatic 107
flight 2, 32
flora 1, 27
Florence, 5
Florida 34
flow
– cytometry 13, 46
– rate 36, 38, 46, 47, 51, 54
flower 3, 5, 17, 21
flowering 5, 17, 18, 20, 22
flowmeter 51, 54, 126
fluorescence 13, 46,
fly agaric 107
fly ash 115
focus 60, 61, 65, 66, 67, 80, 91, 94
Fomes 107
food processing 15
foot and mouth disease 3, 30
forensics 14
Fraxinus angustifolia 101
frogs 2
fuchsin 57, 118
Fuligo septica 109
fume cupboard 52, 53, 117
Funaria hygrometrica 113
fungicide 15
fungus foray 31
fungus, fungi XIII, 1, 2, 3, 5, 6, 8, 9, 11, 13, 16, 17, 21, 22, 27, 29, 30, 31, 91, 92, 93, 94, 107
Fusariella 32
Fusarium graminiarum 113
F. solani 33

Ganoderma applanatum 107
garrya 97
Garrya elliptica 97
Garryaceae 97
gasses 4

gelatine 8, 118
Gelvatol 43, 56, 57, 58, 117, 118
genetic diversity 12, 15
genetically modified material 19
geographical location 17, 19
Germany 8
germination 22
giant puff-ball 107
glass rod 37, 56, 57, 117
glistening ink cap 107
Gloeocapsa 77, 92, 113
glycerine jelly, 36, 43, 51, 56, 57, 58, 118
glycerol 117, 118
GM crops 14, 19
GM pollen 11, 15
golden rod 101
gradient 9, 20, 23, 24, 31, 41, 126
–, concentration 31
–, disease 9, 23, 24
–, dispersal 20, 23
grain 24, 28, 29
– silos 29
Gramineae 97
grass 3, 14, 18, 19, 75, 77, 84, 86, 92, 97
– pollen season 18
graticule 63, 69, 70, 71, 72, 80, 86, 88, 91, 119
– scale 63, 70, 71
gravitational sedimentation 35
gravity 4, 20, 21, 62
Gregory XIII, 1, 4, 9, 10, 11, 15, 91
grey blight of tea 111
groundsel 101
growth stage 17
gulmohor 99
gust 16, 18, 37, 126
Gymnopilus junonius 107
G. penitrans 107
gymnosperms 3, 17, 20, 126

H5N1-strain of avian flu
habitat 15, 17, 21, 73
half-life of spores 22
harvest 27, 32, 77
Hawaii 32
hay 24, 27, 28
– fever 8, 15, 24
hazel 99
head rot 22
health
– health and safety 12, 49, 94
– hazards 11, 12, 24, 25, 28, 29, 30, 47, 79, 91

heart's tongue 103
heather 101
Helicomyces 111
Heliocarpus Americana 101
Helminthosporium 113
Helvella crispa 105
Hesse 8
Heterobasidion annosum 22, 107
Hexane 42, 52, 53, 57, 117
high power 66, 67
higher altitudes 32
Hippocrates 4
histology preparation 91
Histoplasma capsulatum 111
histoplasmosis 111
Holcus lanatus 31
home 25, 28, 94
honey fungus 107
horizontal slide 61, 68
horn of plenty 107
hornbeam 99
horse manure 29
horses 26
horsetail spore 18, 103
host 17, 33, 34
hourly
– concentrations 81
– counts 18, 83, 84
house dust mite 26, 93
human XIII, 1, 3, 4, 9, 21, 26, 29, 30, 39, 46, 50, 59, 111
Humaria granulate 105
Humicola lanuginose 111
H. stellata 111
humidity 17, 20
humming birds
hurricanes 34
hyaline spores, 10, 11, 31, 33, 57, 80, 92, 105, 126
hydration level 20, 22
hydrophobic surfaces, 31
hydrophobicity 21
hygroscopic movements 16
hypersensitivity reaction 29
hyphal fragments 77, 115
Hypholoma fascicularia 107
H. hydophilum 107
Hypoxylon coccinium 105
H. multiforme 105

identification
- of airborne particles XIII, 12, 35, 57, 62, 80, 91-115
-, visual 12, 35
image analysis 13
immunoassay 46, 126
immunological 12, 13, 35, 36, 38, 45, 46
- techniques 13, 36, 45
immunology 127
impact (effect) XIII, 12, 21, 30, 127
impact jet 45
impaction 11, 13, 21, 34, 35, 36, 37, 45, 94 127
- efficiency 21, 37
imperfect stage 127
Imperial College 4, 8, 10
impinge 37
incident drop 33
incubation 23, 39
India XIII, 11, 97, 99, 101, 105, 109, 111, 113
indoor
- air quality 2
- environment 12
indoors 1, 3, 15, 24, 26, 35, 94
industrial injuries 27
inertial
- impaction 35
- spectrometer 47
infection 4, 5, 9, 12, 17, 21, 22, 23, 24, 29, 30, 34, 94
infectious diseases 5
influenza 29, 30
infra red (*see* near infra red) 23, 24
inhalation 8, 27, 29, 30, 50, 91
inhaled
- air 4
- biological particles 1
- fungus spores 8
inoculum 23, 33, 34, 50
Inocybe geophylla 107
inorganic particles 2, 4, 53, 93
insect 2, 3, 5, 9, 17, 18, 22, 42, 51, 56, 79, 92, 93, 115
- compound eye 115
- hairs 79, 115
- leg 115
- pollinated 17, 126
- scale 79, 115
- vector 22
International Association for Aerobiology 14, 120
intramural 1, 9
invertebrate 2, 3

isokinetic efficiency 10

Japanese cedar 18
Juglandaceae 97
Juglans regia.97
Juncaceae 97
juniper 103
Juniperus communis 103
Kalanchoe 99
Knight 5
Koch 5
Koch's nutrient gelatine 8
Koch's postulates 6
Koelrueter 5
Kyllingia 97

label 43, 51, 55, 56
-, plant 42, 52, 53, 57
Laccaria amethystine 107
Lacrymaria velutina 107
Lactarius blennius 107
L. rufus 107
lactic acid 117
lactophenol 117
lakes 30, 93
landfill sites 29
late summer asthma 27, 105
Latin names XV
Leeuwenhoek 5
Legionella 30
legionnaire's disease 30
Leguminosae 99
Leicester 19
lense 5, 59, 60, 62, 64, 65, 66, 67, 88
Leocarpus fragilis 109
Lepiota racodes 107
Leptosphaeria maculans 17, 22, 27, 79, 105, 111
liberation *see*: release 127
lichen 2, 3, 92, 105, 113, 127
- soredium 113
lifecycles 93
Ligustrum ovalifolium 101
lime 101
Lincolnshire 28
Lindenbergia indica 101
ling 101
liquid 13, 21, 31, 36, 44, 45, 46, 69
liquid films 33
London 4, 8
longitudinal traverse 81, 83
Lucretius 4, 8

luminometer 13
lung 24, 25, 27, 28, 30, 39, 91
Luzula campestris 97
Lycoperdon giganteum 107
L. perlatum 107
Lycopodium 9, 51, 53, 55, 103

MAARA 14, 120
macrofungi 92, 127
magnesium silicate 115
maize 11, 20
Malaysia 97
man (see: human) 111
mastic tree 101
Mauna Loa Observatory 32
meadow rue 99
mechanical activity 15, 18
mechanical counters 80
media 25, 27, 35, 38, 39, 57
medical mycology 9, 11
medical
– practices 25
– profession 11
medium
–, collection 36, 46
–, nutrient 5, 8, 35, 36
Meier 9, 32
Melampsoridiun betulinum 105
Melanospora zamiae 105
Melastomaceae 101
Meliaceae 101
Melinex tape 43, 51, 52, 54, 56, 85, 118
Memnoniella echinata 111
Merulius lacrymans 11, 107
mesophilic 27, 127
metabolites 2, 24
meta-xylem 115
meteorological
– data 81
– events 10
meteorology 6
Micheli 5
Miconia 101
microbes, microorganisms 5, 6, 8, 15, 24, 35, 38, 94, 120, 127
microbial
– contamination 13, 15, 44
– content of the atmosphere 6
– life 9
microbiology XIII
microclimate 3

microenvironments 15
micrometer 63, 64, 69, 70, 71, 72, 94, 119, 127
Micromonospora vulgaris 27
Micronectriella nivalis 105
microphotograph 73
microscope 49, 56, 59, 60, 61, 62, 64, 65, 66, 67, 68, 69, 70, 71, 80, 81, 84, 85, 86, 88, 89, 91, 93, 94
–, binocular 60, 94
–, monocular 59, 84
–, structure 59
microscopic XIII, 5, 12, 38, 71, 91
– particles XIII, 5, 91, 92
microscopy 36, 40, 44, 45, 46, 47, 68, 91
mildew, downy 109
mildew, powdery 12, 16, 17, 22, 111
Mimosa arenosa 99
M. pudica 99
miniature wind tunnel 17
Miquel 6, 8
miscellaneous 115
mist droplets 75
mite 2, 18, 26, 79, 93, 115
– leg 115
– pellet 79, 115
mitosporic 94, 111, 127
mobile laboratory 43
model
–, dispersal 18, 20
–, dispersal gradients 23
–, empirical 20
–, exponential 20
–, Gaussian plume 20
–, gradient transfer theory 20
–, K theory 20
–, physical 20
–, power law 20
–, rain splash 33
–, random walk 20
molecular techniques XV, 12, 35, 36, 45, 46
mollusc 2
monitoring
– spore release, 23
– the air, 14
monocotyledons 94, 97, 127
Moraceae 99
moss 2, 4, 5, 18, 33, 92, 113, 127
– spores 113
moth scale 115
motor 11, 39, 43, 119
mould 5, 8, 27

mouldy hay 28
mountain ash 99
mountant 43, 47, 56, 65, 117, 118, 127
moving atoms in the air 4
Moviol 117
mucilage 31, 33
   Mucor spinosus 109
mucus 2, 3, 29
mugwort 101
mulberry rust 105
mumps 30
mushroom 29, 79, 92, 107
– compost 29
– farm 79
Mycena crocata 107
M. inclanata 107
Mycobacterium tuberculosis 30
mycoplasmas 22
Mycosphaerella capsellae 105, 111
M. musicola 22
Myrtaceae 101
myxomycete 127
– spores 109

narcissi 9
narrow leaved ash 101
National Pollen and Aerobiology Research Unit 13, 120
neam 101
near-infra-red 23, 24
Nectria cinnabarina 105, 113
needle 43, 51, 53, 55, 57, 94
nematode 2, 3, 18, 127
Neobulgaria pura 105
Neozygites fresenii 109
nettle 19, 99
New Zealand 31, 93
night 2, 10, 27, 32
Nigrospora 77, 113
nitrogen 2
Nolanea staurospora 107
nomenclature XI
non-turbulent air 36
North Sea 32
nose 39

oat 109
objectives 59, 61, 64, 65, 66, 67, 68, 70, 71, 72, 74, 75, 77, 79, 80, 85, 88, 89
–, magnification 61, 75, 77, 79, 121 122, 123
Observatoire Montsouris 6

occupational lung disease 28
Oculimacula yallundae 34
Oidium 111
oil immersion 61, 67, 69, 94
oil oalm 97
oilseed rape 17, 18, 20, 31, 75, 77, 79, 99
Olea europaea 101
Oleaceae 101
olive 101
Oncopodiella 113
oospore 109
operculum 18
Ophiobolus graminis 105
optical 13, 46
organisms XV
orifice 8, 10, 36, 43, 51, 53, 54, 82, 84, 127
–, constriction 54
outdoor air 1, 2, 3, 6, 12, 24, 83, 93
– – sampling 8, 11, 35, 36, 41
outdoor wind speeds 10
outer frictional turbulence layer 3
outer space 2, 4
oxygen 2
oyster mushroom 107

paddy field 11
Paecilomyces varioti 111
paintings XIII, 91, 93, 94
palm 97
Palmae 97
Panaeolus foenisecii 107
P. sphinctrinus 107
Papularia 113
parachute 20
paraffin wax 42, 51, 52, 53, 117
Parc Montesouris 8
Parietaria diffusa 99
Paris, 5, 6, 8
parkland 31
Parthenium hysterophorus 101
particle XIII, 1, 2, 4, 9, 12, 13, 18, 19, 20, 21, 22, 31, 35, 36, 37, 40, 41, 42, 45, 46, 73, 75, 81, 89, 91, 94
–, biological XIII, 1, 3, 11, 32, 35, 36, 91, 93
–, cloud 18
–, dispersion 2, 19, 31
–, measurement 18, 71, 73, 85, 88, 89, 94
–, shape 1, 16, 19, 20, 21, 33, 91, 93
–, size 9, 13, 36, 37, 40, 46
–, surface texture 19, 20
Paspalum dilatum 97

Passeriniella 105
passive release 16, 17
Pasteur 5
pathogen 10, 17, 30, 33
pathogenic 29, 94, 127
pawpaw 101
PCR 12, 35, 127
peanut leaf spot 111
– rust 105
peat 92, 93
Penicillium 12, 28, 79, 92, 111
P. chrysogenum 111
P. frequentens 28, 79, 111
P. marneffei 111
P. roqueforti 12
perfect stage 127
perforated zinc drum 28
periodicity 10, 17, 30
Peronosclerospora sorghi 109
Peronospora 16, 77, 109
P. parasitica 109
Pestalotiopsis theae 111
Petri dish 9, 12, 36, 38, 39
petroleum jelly 36, 40, 42, 50, 51, 52, 53, 57, 117
Phaeosphaeria fuchlii 105
P. nigrans 105
Phallus impudicans 107
pharmacological 35
phase contrast objectives 61
phenol 117, 118
Philedelphia
Phleum pratense 92, 97
Phoenix sylvestris 97
Pholiota penitrans 107
P. squarrosa 107
Phoma lingam 105, 111
photographs 65, 73, 91, 93, 94
Phyllanthus virgatus 101
Phyllitis scolopendrium 103
physical properties 9
physics 1
Phytophthora infestans 22, 23, 24, 46, 109
pickup balers 27
Pilobolus kleinii 16
Pinaceae 103
pine 75
Pinus 20, 103
P. sylvestris 103
Pistacia lentiscus 101
Pithomyces chartarum 31, 75, 77, 111
P. maydicus 111

plane 99
plant 1, 2, 6, 9, 15, 16, 17, 20, 21, 22, 32, 33, 42, 83, 92, 93
– diseases 22-24
– hair 115
– pathogen XIII, 10, 11, 12, 17, 50
– pollination 20, 21
Plantaginaceae 101
Plantago coronopus 101
P. lanceolata 101
Platanaceae 99
Platanus 99
Pleospora herbarum 105
Pleurotus ostreatus 107
Pluteus cervinus 107
pneumonia 29, 30
Podocarpaceae 103
Podocarpus neriifolia 103
point of impact 33,
pollen - XIII, 1, 2, 3, 5, 6, 8, 9, 10, 13, 15, 21, 24, 26, 32, 33, 43, 46, 50, 56, 57, 59, 73, 75, 77, 80, 81, 84, 85, 86, 88, 89, 92, 93, 94, 97, 99, 103, 118, 120, 127
– concentrations 19, 20
– counts 11, 25, 49, 82, 121
– dispersal 1, 9, 11, 13, 19, 20, 31
– grain 22, 62, 67, 71, 91
– monitoring 14
– peak 14, 17, 32
– release 3, 17, 18, 19
– season, grass 18, 88
– shape 1, 20, 21
– calendar 26
pollination 19, 20, 21
pollinators 3
pollinosis 25, 26
Polygonaceae 99
polymerase chain reaction 12
Polythrincium trifolii 111
polyvinyl alcohol 117
Portugal 79, 101
potato
– blight 10, 22, 23, 24, 46, 109
– virus 9
power 13, 20, 41, 42, 43, 47, 50, 61, 63, 66, 67, 68, 70, 71, 85
primary conidium 109
primary foci 24
Pringsheimia 105
privet 101
propagule 33

protist 2, 127
protozoa 1, 2, 5, 18
pro-xylem 77, 93, 115
Pseudocercosporella capsellae 105, 111
P. herpotricoides 34
Psilocybe 107
Pteridium aquilinum 30, 77, 103
pteridophyte spores 30
Puccinia arachidis 105
P. graminis 105
P. melanocephala 32
P. striiformis 21, 79, 105
pump 8, 11, 28, 38, 43, 46, 47, 54
Pyrenopeziza brassicae 21, 75, 105, 111
Pyricularia oryzae 111
Pyronema confluens 105

Q fever 30
Quercus robur 99
Q. suber 99

rain 4, 5, 10, 15, 16, 17, 18, 21, 23, 31, 33, 34, 41, 43, 44, 45, 92
– forest 92, 103
– shield
– splash 15, 16, 22, 33
– tower 33
raindrop 21, 36
rainfall, 16, 17, 33
rainy weather 8
Ranunculaceae 99
rapids 18
recycling 29
red-cracked Boletus 107
refractive index 67
refuse dumps 29
regional background 19
relative humidity 17, 20, 128
release *see*: liberation 3, 5, 7, 9, 12, 15, 16, 17, 18, 21, 22, 23, 24, 25, 27, 29, 30, 31, 32, 33, 34, 35
– mechanisms 16
–, spore 5, 6, 7, 9, 16, 17, 18, 22, 23, 27, 29, 31
–, pollen 3, 17, 18, 19
–, active 18
respirators 29
respiratory
– allergy 8, 24
– disease 25, 26, 27, 28, 29, 30
– tract 39, 46
Reticularia 109
rhinitis 25, 26

Rhizopus nigricans 109
Rhynchosporium secalis 92, 111
ribwort 101
rice 109, 111, 113
– leaf spot 111
risk assessment of GM pollen 15
Rosaceae 99
Rosellinia aquila 105
rotation speed 40, 89
Rothamsted XIV, XV, 9, 10, 15, 27, 30, 33, 40, 75, 77, 79
rubber particles 77
rufus milk-cap 107
Rumex acetosa 99
R. crispus 99
Rungia pectinata 101
run-off water 33
Russula ochroleuca 107
R. vesca 107
rusts 10, 21, 22, 31, 32, 77, 92, 105
rye 111

S. E. Asia 32
Sacharopolyspora rectevergilla 28, 115
safety 12, 49, 53, 94
safranin 57, 75, 80, 118
Sahara 4
Salicaceae 97
saliva 3, 6, 10, 12, 35, 36, 39, 43, 44, 45, 46, 47, 49, 55, 56, 86, 89
Salix caprea 97
salt bush 99
salt cedar 101
Sambucus nigra 101
sampling 9, 11, 12, 15, 21, 28, 31, 32, 35, 36, 38, 40, 42, 43, 45, 47, 49, 50, 54, 57, 73, 75, 89, 128
–, continuous 45, 50
– devices *see*: air samplers
– efficiency 10, 21, 35, 36, 37, 40, 42, 44, 45, 46, 47
–, isokinetic 50
– rate 36, 37, 38, 40, 45, 47, 50
sand 4
saprophytic 29
SARS 30
scalpel 42, 43, 51, 55
Scandinavia 93
scanning electron micrographs 93
Schistidium antartica 113
Schmidt 9

Sclerotinia sclerotiorum 16, 22, 79, 105
Scopulariopsis brevicaulis 111
Scots pine 103
Scrophulariaceae 101
sea 32
– foam 18
Seale Hayne College 9
season 2, 3, 14, 18, 24, 35, 51, 83, 88
seasonal changes 14, 17, 26, 29
sedge 97
sedimentation 21, 35, 36, 37, 45, 128
seed 23, 92
– production 19
Selaginella pulcherrina 103
Senecio vulgaris 101
Septoria tritici 111
Serpula lacrymans 11, 107
settle 8, 35, 36, 45, 69
severe acute respiratory syndrome 30
shaggy Pholiota 107
sheep 31, 111
sigatoka of banana 22
silicone grease 36, 40, 51
silk threads 20
silos 29
silver birch 99
silver leaf 107
Singapore XIII, 97, 99, 103, 105, 111, 113
siting the trap 49
size fractions 24
skin 2, 3, 27, 79, 115, 117
– scales 79, 115
– test 27
slide (microscope) 6, 8, 9, 10, 11, 31, 32, 33, 35, 36, 37, 40, 42, 43, 44, 45, 50, 51, 52, 55, 56, 57, 58, 59, 60, 61, 63, 64, 65, 66, 67, 68, 69, 70, 73, 75, 77, 79, 81, 82, 83, 84, 85, 86, 87, 89, 93, 94, 122
– box 51, 58
– preparations 63, 91
slimy milk-cap 107
small wind tunnel 9, 27
smog 2
smoke 2, 4, 18, 19
– particles 4
smut 92, 109
snow 33
soil 8, 18, 22, 23, 77, 93, 115
– particles 77, 115
soilborne 22
Solidago 101

sooty bark of sycamore 113
Sorbus aucuparia 99
Sordaria fimicola 16, 105
sore throats 30
sorghum 109, 113
sorrel 99
source 2, 5, 9, 11, 12, 15, 17, 18, 19, 20, 23, 24, 27, 30, 34
–, area 15, 24
–, line 9
–, point 9, 24
South America 33
South Orkney Islands 33
Southern England 4
Spaerutina 105
Spain XIII, 18, 97, 99, 101
Sparganiaceae 97
Spargarium erectum 97
specimens 71, 91 94
spectrometer 47
spectroscopy 13
speed
–, airflow 36
–, fall 4, 20
–, rotation 39, 40, 41, 42, 43, 54, 89
–, wind 2, 9, 10, 23, 28, 37, 50, 83
speed airflow 36
Spegazzinia deightonii 113
S. lobulata 77, 113
S. tessartha 77, 113
Sphacelia sorghii 113
Sphacelotheca cruenta 109
S. reiliana 109
S. sorghi 109
Sphagnum 18
spider 18, 20
Spilocaea pomi 105, 111
splash 3, 9, 15, 16, 18, 22, 33, 34
– dispersal 9, 17, 21, 33, 34
spontaneous generation 4, 5
sporangia dehiscence 18
sporangiospore 109
sporangium 109
spore 1, 10, 11, 12, 14, 15, 16, 17, 18, 19, 20, 21, 22, 23, 24, 25, 28, 29, 30, 31, 32, 33, 35, 36, 38, 40, 41, 43, 45, 49, 50, 54, 62, 67, 68, 71, 73, 75, 79, 80, 83, 84, 85, 86, 87, 88, 91, 92, 93, 103, 109, 113, 117, 118, 121, 122, 123, 128
–, actinomycete 20, 24, 27, 28, 29, 79, 80, 92, 115
– capsule 18, 92
– dispersal 4, 8, 9, 16, 18, 19, 31, 33

–, fern 2, 5, 18, 30, 77, 92, 93, 103
–, fungal 1, 2, 9, 13, 14, 16, 17, 18, 21, 24, 26, 27, 29, 32, 33, 49, 57, 62, 80, 84, 86, 87, 88, 92, 93, 94, 105, 107, 109, 111
– liberation mechanisms 16
–, moss 2, 5, 18, 92, 93, 113
– release 5, 6, 7, 9, 16, 17, 18, 22, 23, 27, 29, 31
– shape 1, 16, 20, 21, 33
– laden dust 24
– suspension 33
Sporobolomyces 17, 32, 75, 111
Sporormia 105
Sporotrichum 111
sporulation 17
spotted tough-shank 107
Sprengel 5
St Mary's hospital, Paddington 11
Stachybotris 111
stack burn 111
stage 4, 17, 18, 22, 37, 38, 39, 59, 61, 62, 63, 64, 65, 66, 66, 67, 68, 69, 70, 71, 72, 79, 85, 94
– micrometer 63, 64, 69, 70, 71, 72, 94, 128
stain 13, 43, 46, 57, 75, 79, 80, 118,
Staphylococcus aureus 115
stem canker 17, 22, 79, 105
Stemphylium 113
Stenochlaena palustris 103
Stepanov 8, 9
Stereum 107
stigma 21
stink horn 107
Stoke's Law 36
storage 28, 51, 53
stored products 24, 28, 38
stratosphere 2, 3
straw 24, 29, 30
Streptococcus 30
Streptomyces 115
stroboscope 41, 128
Stropharia aeruginosa 107
suberosis 28, 79, 111
substage structures 59
sugar beet 18
sugar cane 32
sulphur tuft 107
sunflowers 23, 79
Surpula lacrymans 11, 107
Sutton 9
Sweden XIII, 33, 97, 99, 103
sweet chestnut 99
sycamore 101, 113

take-all of wheat 105
take-off 12, 15
talc 115
Tamaricaceae 101
Tamarix pentandra 101
Tamonea 101
tape 12, 39, 40, 42, 43, 44, 49, 51, 52, 53, 54, 55
Tapesia yallundae 34
Taraxacum 101
T. officinale 20
Taxaceae 103
taxonomy XV
Taxus baccata 103
tea 111
teleomorph 128
teleutospore, 108, 128
temperature 2, 22, 27, 36, 56, 57
template 54, 55
terminal velocity 9, 19, 20, 21, 33, 36, 128
Tetramitus 115
Tetraploa aristata 111
Texas 32
Thalictrum 99
the blusher 107
The Hague 14
Thecamoeba 115
Thelephora terrestris 107
thermal convection 19
thermals 2
Thermoactinomyces 115
T. sacchari 28
thermophilic 27, 29, 128
Thermopolyspora polyspora 27
thresher 28
thrips 93
thunderstorm 84
Tilia 101
Tiliaceae 101
Tilletia 109
T. barclayana 109
T. caries 109
T. holci 109
Tilletiopsis 17, 75, 111
timothy grass 97
Tolyposporium ehrenbergii 109
tonsillitis 30
topography 73
tornadoes 2, 4
Torula 113
T. herbarum 113

toxic metabolites 24
toxicity 35
toxicoses 24
toxin 2, 13, 47
trace 51, 53, 54, 55, 56, 58, 80, 81, 82, 83, 84, 85, 86, 87, 128
– width 80, 81, 83
trachea 39
Trachycarpus fortunei 97
transmitted light 59
transport 1, 2, 5, 9, 12, 26, 31, 32
– by wind 9
transverse traverse 81, 84, 87
trap 9, 10, 11, 13, 15, 17, 19, 27, 30, 31, 32, 35, 36, 37, 38, 39, 40, 41, 42, 43, 44, 49, 50, 51, 53, 54, 55, 57, 58, 68, 73, 74, 75, 76, 77, 79, 81, 82, 83, 85, 86, 87, 89, 90, 93, 118, 119, 120
trap shape 37, 39
trapping XIII, 8, 12, 17, 21, 35, 37, 40, 49, 50, 51, 53, 82, 84, 87, 89, 117
– devices *see*: air samplers
– notebook 51
– surface 40, 49, 50, 51, 53, 82, 87
travel trap 43, 44, 77
traverse 80, 81, 83, 84, 85, 86, 87, 88, 89, 90, 128
– width 80, 81, 86, 87, 88, 89
tree pollen 14
trees 107
triangle leaf bursage 101
Trichoconis padwickii 111
Trichoderma viride 111
Tricholoma album 107
T. nudum 107
Trichothecium roseum 111
Triphragmium ulmariae 105
Tristan da Chuna 115
Triticum sativum 97
troposphere 2
trypan blue 57, 79, 118
Tubaria furfuracea 107
Tubercularia vulgaris 113
tuberculosis 6, 30
turbulence 2, 9, 16, 18, 19, 20, 36
turbulent
– air 17, 36
– weather 32, 33
turgor pressure 17
tyre roll 115

UK XIII, 13, 14, 18, 30, 62, 75, 77, 79, 84, 97, 99, 101, 103, 105, 107, 109, 111, 113, 115

Ulmaceae 99
Ulmus glabra 99
umbellifer pollen 77
Umbelliferae 101
unit area 33
uredospores 79, 105, 128
Uridinales 105, 128
urine 3
Urocystis agropyri 107
Uromyces fabae 105
Urtica dioica 99
Urticaceae 99
USA XIII, 97, 99, 101, 111
Ustilago 75, 77
U. avenae. 109
uv-light 22

vacuum pump 28
velocity 2, 9, 19, 20, 21, 33, 36
Venezuela XIII, 97, 99, 101
ventilated 29, 52, 53
Venturia inaequalis 105, 111
verdigris agaric 107
vernier scale 61, 68, 85, 85, 128
vertical arms 39
– mast 31
– spread 33
Verticillium dahliae 111
viability 5, 15, 22, 24, 27, 29, 32, 35, 36, 38, 39, 46, 128
virulence of spores 12, 13
virus 3, 9, 22, 30, 128
volcanically warmed soils 33
volcanoes 4
vortex 44

Walleria sebi 111
walnut 97
weather 2, 24, 35, 51, 73, 84
–, dry 20
– events 18
– forecast 11
–, humid 21, 31
–, rainy 8
–, turbulent 32, 33
weeping widow 107
West Africa 32
wet-dry cycles 17
wetness sensor 17
wheat 17, 22, 79, 97, 105, 109, 111
whirlwinds 4

white rust 109
whooping cough 30
willow 97
willow herb 20
wilt of plants 111
wind 1, 2, 3, 4, 5, 6, 8, 9, 10, 17, 18, 22, 23, 27, 28, 31, 32, 33, 34, 37, 42, 43, 44, 45, 50, 54, 83, 92, 128
- dispersal 4, 31
- pollination 5, 125
- speed 2, 9, 10, 23, 28, 37, 50, 83
- tunnel 9, 17, 27, 28, 128
- vane 10, 43, 44, 54
winged pollen 20
wood blewit 107
wood
- chips 24
- decay 107
Worcester 75, 79
work environments 83
work place 28, 94

Xanthomaonas axonopodis 22, 34
Xanthoria parietina 92, 105
Xylaria polymorpha 105
yew 103
Yucaton 32
one of infection 23

zoonotic disease 30